# NO
# BACKUP
## A SETUP AND MORE

The Accomplishments
of a Good Fairy

Malcolm Crockett

Published by BN Publishing Services

Cover design by BN Publishing Services

ISBN: Printed in the United States

**PUBLISHING**
S E R V I C E S

# PRAISE FOR

A truly Canadian story laced with international flavours, Malcolm Crockett's engaging autobiographical novel manages to combine humour with a serious look at a number of diverse subjects, including anti-nuclear activism, workplace prejudice and the darker side of social work. Readers of W.O. Mitchell and Mordecai Richler will enjoy Crockett's childhood stories from a small Vancouver Island community in the 1960s that feature such characters as the scoundrel Jake-Jack-John and the one-of-a-kind Eight Bells. Follow Crockett's protagonist through life's ups and downs, encountering the famous and the near-famous and journeying through self-discovery in this charming and informative tale. **Catherine Marie Gilbert, author and historian.**

Be prepared for a wild ride when you engage with this work. Malcolm Crockett's 'auto ethnographic novel' brings the reader into the midst of the deep and incessant homophobia that continues to plague too many social spaces in what purports to be tolerant at least, accepting at best, Canada. Taking seriously his union saviour's advice to write his way to survival, Crockett drops us into aspects of a life one could never anticipate. From Maori elders treating his childhood comas, to the riches of growing up on Vancouver Island from leftist activism to the vagaries of government employment, surprises await us at every turn. **Celia Haig-Brown, Phd, FRSC, Professor Emerita, York University, Toronto. Author of Tsqelmucwilc: The Kamloops Indian Residential School ~~ Resistance and A Reckoning.**

# MORE PRAISE

As Malcolm Crockett's former practicum student I am thankful to have learned from a structural social work mentor who helped shape my clinical foundation. *No Backup* demonstrates his artistic use of words to connect social location, epistemic privilege and worldview to resiliency and advocacy. This narrative inspires intrapersonal praxis and reflects the sharing of knowledge between interpersonal generations of clinical practice. A must-read for anyone contemplating or engaging in the challenges and accomplishments of social work. **Christine Woltman, BSW, MSW, RSW.**

In No Backup, Malcolm Crockett brings together his personal experiences and many influential cultural anecdotes that took place over a number of years. He is able to also reference some of the changing cultural norms and provide insight both into what it was like both to grow up in a small town on Vancouver Island and into some of the political situations that influenced his life. **-- Jean Crowder, former NDP Member of Parliament, Nanaimo Cowichan.**

# DEDICATION

In loving memory of Maggie Kennedy Garrick, Artist and
Anthropologist and her son, David Garrick, Anthropologist and
Earth Scientist.

# CONTENTS

# About the Author

Malcolm Crockett is a graduate of UBC and Carleton University. He has worked in New Zealand, Australia and South East Asia and also in a professional capacity within government for over 30 years. He holds a Field Practice Supervision Certificate from the University of Victoria and has mentored numerous Social Work students while working in government and as President of the Cowichan Neighbourhood House. He was a member of the BC Association of Neighbourhood Houses delegation to the 90th Anniversary founding of the International Association of Neighbourhood Houses in Berlin, Germany and is an Honorary Life Member of the BC General Employees Union. Mr. Crockett lives in the Cowichan Valley on Vancouver Island where he is a Board Member and docent of the Crofton Old School Museum.

# About the Book

From the late 1960s until the early 1990s the Canadian government, assisted by the RCMP, conducted a comprehensive program of harassment of Gays and Lesbians in the Canadian Armed Forces and the Public Service. Several men and women targets of systemic intimidation committed suicide.

When RCMP officers involved were stationed in other parts of the country, homophobic harassment continued. In **No Backup: A Set Up and More**, the author describes working as a young Gay Social Worker when RCMP Officers informed him that he would not receive "Backup' as the force" ... didn't want anyone in the detachment or the community liking him too much."

A Union Executive recommended Shop Steward training and encouraged the writer to document his early life experiences, including his sibling rank and the life experiences where he grew up. Fascinating characters and dynamic adventures emerge.

From surviving two viral meningitis comas as a child in New Zealand to serving on a global planning committee organizing anti-nuclear protest to lobbying the Canadian Prime Minister to rescue ten people from the Pinochet dictatorship in Chile, to surviving a life-threatening call-out where RCMP officers placed him in a room with a severely impaired homicidal and suicidal adult, the protagonist documents the activities and the unique influences of hometown characters. A remarkable tale of survival and success.

# PREFACE

… every autobiography is a work of fiction and every work of fiction
an autobiography

Time to be Earnest, A Fragment of Autobiography P.D. James.[1]

# CHAPTER 1

My friend, colleague and shop steward, Nancy, warned me at least twice, but likely several times, "Whenever an initial appraisal is complete, he will say or do something that sends the worker out of his office in tears. Be prepared. Come out of his office looking like you've won."

In February 1986, I met with a human resources executive from the Public Service Commission to receive a six-month job performance evaluation. I was supposed to receive the evaluation in November, but this person kept sending me three-part memos, a particular form of office communication in triplicate, to request a postponement.

First, he had a regional meeting; next, he "just plum forgot and double-booked for the scheduled time." In January, he assured me he would "get it done within a month."

I didn't realize at the time that the postponements were part of a micromanaged strategy to ensure that two years after my May 1985 start date, it would be another nine months before I could apply for work in another location. One could apply to work elsewhere in the Public Service only after two years since the last successful evaluation.

During the session, after a positive assessment of my social-work skills, the executive and I signed two copies of the evaluation form using a fancy pen from his plastic envelope in his shirt pocket.

The signing procedure was significant. I had passed my initial six-month probation period.

He was short, scrawny, and sinewy with a pockmarked face and scarring left over from adolescent acne. His worst feature was his hair, which he kept shaved above his ears, leaving the rest thinning, crinkly, and a colour of dark blond or light auburn that could only be described as *shit-brindle*.

He oiled, or greased, what remained, enhancing the nondescript colour, and he combed the strands across his head to ensure that some stuck to his forehead. If there were a literary character he may resemble, it would be the antithesis of a *Mr. Darcy*. He was more like a walking, talking, nightmarish reincarnation of *Gollum*.

After we'd both signed the evaluation forms, he did something curious and inconsistent with his otherwise taciturn manner. He took the pen from me and, rotating his wrist, moved it as though swinging a lariat. This was out of character for someone who distinguished his actions and communications with a distinctly laconic economy of movement or expression.

He put down the pen, folded his hands on his lap, then smiled and asked, "Are you a homosexual?"

The question warranted a measured response. Everyone in town knew I was gay. Why was this conservative evangelical asking?

I raised my eyebrows. "Have I ever said or done anything to indicate I'm not?"

"Just answer the question."

"Have I ever said I wasn't?

"Are you a homosexual?"

"Who wants to know?"

"I want to know. Are you a homosexual?"

"Why do you want to know?"

"I just need to *know.*"

"Are you asking on behalf of Personnel?" He interrupted. "No, I *just* need to know."

"Why?"

"C'mon. Just tell me."

I should have walked out and found Nancy. I should have asked, "What planet are you from?" Instead, I said, "I make it a particular point of not acting like John Wayne, but if Tina Turner ever said, 'Honey, I'd never kick you out of bed,' I'd stick around. What about you? You wouldn't turn down a roll in the hay with Mick Jagger or David Bowie?"

"Are you saying you're bisexual?"

"I wasn't saying any such thing. Now what about you and the Rolling Stones?"

His tone instantly changed to anger.

"I would never encounter the Rolling Stones, and I certainly would not sleep with Mick Jagger or any member of his band or the other fellow you mentioned. But you might. Maybe you have

3

already."

I never dissociate, but in that moment, my mind drifted back unexpectedly to Brisbane, when the Stones arrived in a landau pulled by four white horses. Mick climbed onto the back of one of the horses, waving a tail attached to his tights like a monkey climbing onto the stage.

"What is this about?" I asked. Silence.

I asked again, and he finally decided to respond.

"Two members of the RCMP came around to tell your supervisor they're not going to provide backup for you. They don't want anyone in the detachment or the community thinking any one member likes you too much."

"That's a particularly interesting way of explaining paramilitary homophobia."

"Speak English."

"You, or they, profile paramilitary heterosexism in a rather interesting manner."

"I have no understanding of what you are saying."

"Clarifying question, you're suggesting I'm responsible for how much individual members of the RCMP like me?"

"You do have your wiles."

"What are my wiles?"

"How you dress. How you speak? How you walk."

"What's wrong with how I dress? It's *'Clean for Gene'*[2] on steroids, button-down Oxford cloth, quality sweaters, pressed dress slacks, and brogues." I glanced at the evaluation form. "I see you've rated my attire as 'Adequate.' What else?"

"Are you a homosexual?"

"Have I ever said I wasn't?"

"No."

"So you've had a meeting with the RCMP about my sexual orientation?

"Yes."

"Did they make the appointment in advance?" Silence. "When did this meeting occur?" Silence.

"Did anyone take notes?"

"You're taking this too seriously."

"I'm taking this too seriously? You just told me you've been officially informed that members of the RCMP won't provide backup for me. I have a statutory responsibility to involve them the moment I assess a potential violation of the Criminal Code of Canada. We also have protocols in place for any social worker who has concerns about their own safety or the safety of the public."

Silence.

I recited recent statistics about women recruits who have left the force complaining of sexual harassment.

"Who exactly attended this meeting?" Silence.

"When did the meeting take place?" No response. More silence.

Then, "As I said, you're blowing this out of all proportion."

"Wait a minute. You drag me in here and throw your pen around like you're Miss Montana in a Miss America baton twirling talent competition, and tell me the police aren't going to provide backup for me because they don't want anyone to think someone may *like me too much.* How much is too much?"

Something about this question triggered silent, seething anger.

He stood up and approached, leaning his face to within inches of mine.

"If you were a woman, you'd be considered a slut." My only reaction was a deep breath.

For a moment, I said nothing.

"Let me try a different tack. You said they don't want anyone in the detachment or the community thinking that any one member,"

He interrupted again. "I wouldn't go there if I were you."

"Any one member? Only one?"

He raised his voice, still invading my space with his presence. "As I said, if you were a woman, you would be a slut."

After that, we sat in silence until he said, "When did you first know you were a homosexual?"

Remembering Nancy and others, having said this particular human resources executive completed evaluations by causing anxiety, even tears, I composed myself. Should I mention that the construction of identity involves a process?

Or say I knew a Wobblie, an Industrial Workers of the World hobo who left my hometown with an old tar, a sailor named Eight Bells (which actually is a maritime term signifying the end of a watch) or that the bank manager and an art teacher in my hometown lived together?

"Well . . . ?"

I debated whether to give him what he wanted, a straight answer, so to speak. But what he didn't realize was that my truth had always been layered in code, in camp, in songs.

"The first time I heard Judy Garland sing."

"I was expecting something more specific."

"Walk up the Avenue," I offered.

"What's this 'walk up the avenue'?"

"A song."

"A song?"

"Yes. A song by Irving Berlin in the movie Easter Parade. 'The Vanderbilts are having us to tea/ but we don't have the carfare, no siree. No siree. So, we'll walk down the avenue . . . '

"Why don't you just quit," he said in exasperation. "You've proved you can do this work. You've got a good evaluation. You can just leave."

"No siree," I replied. "I've got a good evaluation. It should have been a much better rating, and you know this. After years of exclusion based on discrimination, I finally got a job with reasonable pay and benefits, and I'm in a union.

In another place, in another time, I could likely do many other things, likely with proficiency, I'm not going to leave this job based on your prejudice, wholly inconsistent with the values of this profession, I might add."

"Are you finished?" he asked me. "I don't know. Am I finished?" No response.

"It may not have been Judy Garland. Discerning identity involves a process; you should know this. It may have been when I heard Barbra Streisand singing *Second Hand Rose*." I add, "Lyrics by Grant Clark, and if memory serves, original music by James F. Hanley."

"What is that supposed to mean?

"'Even Jake the plumber, he's the man I adore, he had the *noive* to tell me he's been married before…' I want you to remind our good friends in the detachment that I have a statutory responsibility to involve them in assessments and investigations.

That decision is at my discretion. If there *is* a situation where they don't provide backup for me and I'm injured in any way, I'll ensure the matter is brought up in the House of Commons. A parliamentarian who sits on the standing committee responsible for annual appropriations to the RCMP will address it."

Again, he asked me, "Are you finished?

"Are you? "I am."

I paused, recalling Nancy's advice and thinking of Mick Jagger mounting the stage swinging his monkey tail. Only then did I open the door to the hallway. The supervisor of administration services loitered by the copy machine as I caught the enquiring eye of the child-welfare clerk. I smiled at them, then clenched my fist and punched the air above my head.

"Yes! Yes!"

After leaving the office, I met Nancy by the water cooler. "Do we have a grievance?" she asked excitedly.

"It's much more complex."

"Wait here," she said.

Swinging her arms like a soldier on parade, she marched into the district supervisor's office to ask if she could take "fifteen minutes for a union matter," adding, with a smile, "you know any such request cannot be unreasonably refused. That's the language in the master agreement . . . 'cannot be unreasonably refused.'"

We grabbed our coats, as it was well below freezing outside, and walked across the parking lot toward one of the Dodge Diplomat sedans used for official business. They were navy blue with the crest of the Province of British Columbia on the front door.

I gave Nancy a copy of the employee-appraisal form.

After a glance and noting it had been signed, she asked, "So what's there to grieve?"

I explained that the supervisor had said that two RCMP officers had met with him to let him know they weren't going to provide backup for me.

"Just two of them?"

"Who knows? That's what he said."

"Well, you can't grieve that. We don't have sexual orientation language in the master agreement."

"We don't have any sexual orientation human rights legislation in the province or the whole country, for that matter."

"Well, you're just going to have to suck it up."

"Great choice of words, sister. That's what I think they're worried about."

We laughed and acknowledged which individual officers were our allies. We noted the members of the General Investigation Squad who accompanied us on sexual abuse investigations and remained supportive, professional detectives.

"Other than them," Nancy said, "it's a culture of homophobia. Cal, one of the GIS detectives, told me that if a new officer comes to town and he's single, he has two weeks to go out on a date with a woman, preferably a nurse, a teacher, or a social worker. If none is available, a waitress, and, get this, if she's white, has all her teeth, and no members of her family involved in the drug trade."

"I love this work, Nancy. I'm good at it. I love the people. I love being their advocate. I don't want to get beaten up in an intimate family violence call-out. Or worse."

She laughed heartily. "I'd like you to talk with my sister,

Carole-Anne. As I've probably mentioned several times, she's the chair of our Social Education and Health Services local in the Prince George and District Labour Council. Right now, she's acting as the component chairperson of all our provincial locals. She's probably got some ideas on how you're going to survive. Like what you, or maybe the union, can do to ensure you don't get killed or worse. I'll give you her number."

# CHAPTER 2

**Carole-Anne**

I called Carole-Anne and was not surprised Nancy had filled her in. She invited me to spend the weekend at her home and suggested we meet at 5:30 on Friday at Cappuccino and Other Arts, a café in Prince George.

As I had Friday off work, I left town around 8:00 a.m., arriving early and with no idea what Carole-Anne looked like. The café was the hippest place in the North with colourful local funk art on the walls. No sooner had I ordered a latte than I felt a hand on my shoulder.

"Jeremy?"

Carole-Anne was tall and well-dressed. I first noticed her smile.

"I'm so glad you could make it," she said. "Let me pay for your coffee."

"That's okay. I've got it. What would you like to drink?"

"Mineral water, but I'll get it. Grab that table in the corner by the window."

It seemed to take an eternity for Carole-Anne to get her mineral water. Other people who knew her spoke with her while I waited.

When she finished talking, she took a small notepad out of her shoulder bag and wrote while speaking with a grey-haired, academic-looking man.

"Sure, I'll attend," she said, "but if you can send me the details before the meeting, that will be helpful."

As I watched, I saw she was the best-dressed person in the place. She wore a fitted, wine-coloured leather jacket, a multi-coloured sweater, pressed tweed slacks, and a complimentary paisley scarf held in place by a large, round, silver Haida pin.

She was studiously well-groomed, for want of a better description, and she looked professional. She wore a dark shade of lipstick, and her dark-blond hair was short and well-styled. Her beaded earrings were small dreamcatchers, but not so small that they couldn't hold alexandrite-blue Viennese beads and small anthracite arrowheads.

Her earrings were the first things I commented on when she arrived at the table.

"Those earrings are really quite something. Where did you get them?"

She reached her hands up to touch them.

"These? Locally made. I had to check which ones I was wearing. I have a fair few. There's a saying in town, 'you can tell who the feminists are because they wear the wildest earrings.'"

I laughed appreciatively at her humour.

"I'm so glad you've come," she said again. "I'm sure Nancy has told you all kinds of things about me, but it would be most helpful if you let me know what you know about the structure of the union.

At this time, I'm acting as the chairperson, chair-tender would be a more accurate term, of the component executive. Do you know what that involves?"

"I know what the component structure is. Each local is represented on its respective component executive based on a formula that takes into account the number of members. Our local has one representative, but I can't imagine what constitutes the role of the acting chairperson."

"Good response. I'm also the chair of the local association for all social workers, deputy sheriffs, income-assistance workers, mental-health clinicians, and audiologists, and I serve on the Prince George District Labour Council. But my real love is serving on the international board of the Association of Humanist Clinical Counsellors."

"And the component executive?"

"Indeed, a role that takes a lot of my time. It involves seemingly endless consultation. How was the drive down?"

"Hair-raising. I caught the sun reflecting off the icicles through the Pine Pass, and wish I'd brought sunglasses. It was so bright."

"Sometimes, too hot, the eye of heaven shines."[3] "So it seems. That's why I'm here."

"Quick, drink up," she said. "This isn't the best place to talk. Just down the street is the best Greek buffet north of Vancouver.. My treat."

"Only if you'll let me leave the tip."

"Sure, if you want to. I'm really looking forward to hearing a comprehensive profile of what you're going through and what else is going on in your office. And tomorrow afternoon, I've arranged for us to meet with some school-board workers who are in quite a complex conflict with the chair of the school board."

"That's right up my alley."

The Greek restaurant did have the best buffet in Prince George, and we started with taramosalata and finished with a few pieces of sweet, tasty, nutty baklava. Over the various dishes we devoured, including two servings of delicious lamb souvlaki, we talked about everything.

I followed Carole-Anne's car halfway across town, then along several backroads to her cabin in the hills overlooking the river.

After we got the wood stove lit and were sitting in the living room drinking hot chocolate, she said, "I have a book for you, *Strunk and White: The Elements of Style*. I've worked with lots of young social workers and related occupational groups and clients but, and I stress, if you at any time, think I'm treating you like a patient, let me know right away.

What I suggest you do to keep yourself sane is write, every day, about your earliest life experiences, your role and rank in your family, how you dealt with any authority you've ever encountered, and how you survived. This will give you the insight and critical thinking to survive in the toxic environment you find yourself."

I explained that after UBC, I returned to New Zealand, where I'd lived as a child, and that within the New Zealand Intellectually Handicapped Children's Society, I worked with a sociologist who suggested I do something similar.

"As a result, I wrote an epistolary short story and I sent it to my folks. I could get you a copy."

"That's a great start," she replied. "What I'm asking is that you probe the backroads of your mind. You'll see, it will be helpful. The union stuff will just happen. As soon as you've completed basic shop-steward training, you can take courses from the local labour council. That will also give you an anchor group in the community to debrief and strategize with. You'll be fine. You'll survive. Hell, you'll succeed."

\*\*\*

When I got back to Dodge, I phoned my folks, and Dad sent me what I'd written to him years before. Shortly after this, I learned that Carole-Anne had not been elected as the Chairperson of the Component Executive. During the election, executive members who had said they would vote for her voted for someone else. She was devastated and, as a consequence, did not run for the position of Local Chairperson; thus, she left the Component Executive.

When her term on the Labour Council expired, she didn't run for a position as the Local's designate for the Council. When we talked about it, much later, she said, "Everything is a learning experience. I'm a stronger and smarter woman, better able to see what things are important."

Within a couple of months, she left town to be with an artist she knew, a handsome young man a few years younger than she was, as healthy and handsome as he was personable. They married in a small ceremony and moved to a community on the Island.

When my folks sent me the epistolary short story I wrote from New Zealand, I read what I'd written and realized that the practice of writing to survive, to understand, and to process had begun years before. This is what I'd written:

Dear Dad.

The last time I was in Tokoroa, someone shared the following narrative with me. Was it like this?

Old Hine clenched her jaw and fought the urge to shed tears as she recalled the first night she came to the little Pakeha boy's house on Dreghorn Place. "Not all heart, no head," she kept saying to herself as she carried her sack with the big jar of squeezed lemon juice. That night, she and the others would go to tell the boy's dad they had come to bathe him and fan the poor little Pakeha to bring him out of the Sleep of a Thousand Dreams.

The last time they came, Old Hine and the others watched through the window. The little Pakeha lay like he were already dead. The Reverend Anderson came and then went. The doctor came to tell the parents that there's nothing more he can do. The mother cried like her heart was breaking, and the dad sat beside the bed and cried too.

Old Hine and the others gathered in a horseshoe formation outside the window, the little Pakeha lying like he was already dead

17

on the bed. One of the island men from Tonga remembered when the little Pakeha first came to town.

He broke away from his parents to go to where the Māori and the island people stand around everytime he visited the market. First, he looked down at the gritty pumice gravel and kicked it with his foot, then started to sing *Bula Malaya Kae Viti Tie-eng Gah* like a little bird who learns songs.

Sweet voice, happy face, big eyes, pretty, pretty, curly hair.

Where did this little Pakeha learn this? Amelia DeThierry's granny, she knows the family and said he learned it in Fiji. Every time he visited people, he came over to sing and to get us to teach him songs. He lay there that night like he was already gone. Everyone was so sad. Sweet little Pakeha. If he died, it would be an omen. A bad omen.

Old Hine and the others knew the time to sing to his spirit to call him out, or else he would die that night. It was February, and he boy had already been in the Sleep of Seventy Winds for twenty- seven days and twenty-seven nights. Slowly, first the women, then the men, softly sang. *Po KareKare Ana.* First, the women stopped. Then the men.

Maybe the poor little Pakeha would die, and his parents would go back to Canada. Maybe nobody would finish building the pulp mill. No good union jobs.

Knowing this and liking and respecting the family, everyone started to sing again. They sang all night. Next morning, the little Pakeha rolled onto the floor and crawled out onto the yard grass.

"What have you been singing?"

His parents are still asleep on their chairs by his bed. We tell the boy he has been in the Sleep of Seventy Winds. Old Hine held him in her lap and sang. Soon he hummed along with her, and the Reverend Anderson came and saw that he was alive. He said we should sing the Magnificat because it was a miracle.

'My soul proclaims the greatness of the Lord/and my spirit rejoices in God my Saviour.' Then we sang the Gloria and the Magnificat again. 'He has filled the hungry with good things/sent the rich away with empty hands.'

Old Hine knew any hymn sung in English was plainsong, and the boy was good. They took him into the house to eat porridge, which she knew was a good thing, and his mother picked all the gladiolas in the back yard and gave them to us. Then she dug up the bulbs.

I still remember that garden. For years, whenever gladiolas bloomed, I wondered if they had bloomed in Tokoroa too. The memory was part miracle, part warning, not just that I came close to dying, but that survival had witnesses.

"Not to eat. Plant these, and when they bloom, remember this night. Thank-you. Thank-you."

The boy did fine these past months 'til he went to Matta Matta and came back again, sick. Now he has been on the Sleep of a Thousand Dreams," thirty-one nights, thirty-one days.

His mother is sick too, in the hospital in Waikato, then

Auckland. She got the same sick Sleep of a Thousand Dreams as the boy. The dad just sat at the kitchen table with Irish whiskey and tried to write letters on blue aerogrammes, but he couldn't write because he was crying so hard.

That's when Old Hine and the others came. The men went into the house to tell the dad they had come to bathe the boy. The dad sobbed, big eyes so red from crying, and he thought the boy had already gone, that we had come to prepare him for burial.

He went out onto the stairs, and he cried and cried. The men held onto him and helped him sit down. "There, there, Fred. We're here with you until morning." But he got up and went to watch Old Hine and the other women bathe the boy.

We bathed the little Pakeha in lemon juice, first his face, then his hands, his feet, his arms, his legs, then his body. All this time, we fanned him and sang. When the juice dried, we bathed his face again, and when we put the cloths with lemon juice under his nose, he was startled.

We sang louder, keeping the juice away from his eyes, and ran our hands through his pretty, curly hair. All night, we sponged the little Pakeha's body, so thin, so wan. We held him upright and gently pulled each finger, each toe, calling him back from the Sleep of a Thousand Dreams.

The old ones wrestled with him, telling him his mother would be fine too, but that he must come back. "You too must come back."

By morning, the doctor came and said that what we all did would be no good, that it was "just a matter of time." Then his uncle came up all the way from Christchurch, and when he was outside talking with the men, the boy asked, "Who? Who?"

We said, "It's you. You are back from the Sleep of a Thousand Dreams."

"No. I know it's me. Who is that voice outside?"

We told him it was his father's uncle from Christchurch.

"My Great Uncle Fred? Here in Tokoroa? He came here first in 1898. Why is he visiting? Is someone sick?"

We all laughed. The great uncle is a Labour Party man. He told us that before they were Canadian Pakeha, they were Scots-Irish Pakeha. That's why the boy knows how to sing like a bird and learns all our songs. That's how we remember the little Pakeha.

The dad knew. He told us the doctor's medicine couldn't cure his boy. He and the great uncle knew the people brought the boy out of the Sleep of a Thousand Dreams and the Sleep of the Seventy Wind journey. The great uncle opened his pocketbook and gave us all the notes he had, then he took out his change purse and gave away his silver coins.

Reverend Anderson and his wife came around again, and she said, "How grateful is the sound of noble deeds to noble hearts who see but acts of wrong."

Old Hine asked if there was a tune, and Mrs. Anderson said it was Sir Alfred, Lord Tennyson's song, but it had no tune.

All head, no heart, Old Hine thought. Good words for plainsong.

"Too bad," she said.

We told them they must take him to see Guide Rangi in Rotorua before a fortnight has passed. They took him.

When the mother came back, she said it was a combination of the doctor's medicine and how we cared for him, including the singing and the lemon juice bathing. Nobody told her that the doctor did nothing for the boy.

*\*\**

A sociologist I worked with conducted some interviews for me, as she felt it would be more methodologically appropriate if another person inquired about the intervention. Dr. Gillian Phillips was seconded to the Intellectually Handicapped Children's Society from her tenured position at the University of Wellington.

Her project involved "conducting interviews to establish social, environmental, and ecclesiastical determinants of intellectual handicap in the North Island of New Zealand."

Gillian and I met over lunch at the Christopher Park Home for the Intellectually Handicapped, where I was employed as a live- in child-care worker. The first time we met, she was holding court with the other female staff members about the history of women's suffrage in New Zealand.

Methodically, she profiled the organizing efforts of Kate Sheppard, the foremost advocate for women's suffrage, who, in 1887, was the national superintendent of the Women's Christian Temperance Union Franchise Department.

She also discussed how, in 1893, Meri Mangakahia, a Māori activist, led a delegation of Māori women to an assembly of the Kotahitanga, the pan-tribal Māori parliament, to seek the right for women to vote and generally participate in the Kotahitanga.

"In September of 1893, the governor signed into law the electoral Bill that allowed for women's suffrage and in the general election of November of that year, 85 per cent of women over twenty-one were registered to vote, and 82 per cent of those voted.

"It wasn't until 1933 that we elected our first woman to parliament. Anyone remember her name?"

Several of the women staff members said, "Elizabeth McCombs," and the house cook added, "Aye! She stood as MP for Lyttelton."

I asked if anyone would like to hear an "herstorical" anecdote about Canada's first woman Member of Parliament. They all agreed to hearing, and I recounted how, when Conservative and Liberal Members of Parliament harassed Agnes McPhail, a member of the Cooperative Commonwealth Federation about her not wearing a hat, and one male MP asked, "Are you ever worried you may be mistaken for a man?" her response was, "Are you?"

This caught Gillian's attention, and throughout the afternoon, she asked me many questions about what I did at the Intellectually Handicapped Children's Society, how long I had lived in New Zealand, and what I had studied at UBC. When she learned I had lived in Tokoroa as a child and had been in two comas, she asked if I would like to accompany her on her interviews.

I checked with the director, who said, "By all means. I've told the university we will provide Dr. Phillips will any support she requests."

And so, I found myself a witness to numerous long, qualitative interviews with several particularly interesting families.

It was Gillian's thesis that three specific circumstances contributed to the incidence of developmental delay in the New Zealand population: settlement privilege based on ecclesiastical association, determining which faith communities got the best land; the intergenerational legacies of who was related to whom being obscured by immigration and the practices of levirate and sororate among settler populations, and in utero damage caused by the practice of top dressing (crop dusting).

As we drove to various places to conduct interviews, Gillian said several interesting things.

"Those associated with the Church of England, which was the established church, got the best land. Various denominations of Methodists acquired farmlands of lesser value, and, particularly, sects like the Primitive Methodists, Non-Conformists, and Roman Catholics received the worst tracts of land. Presbyterians got the rocky lands of the South Island around Dunedin.

"With respect to the social implications of these settlement patterns, there were various forms of sectarian prejudice. Anglicans couldn't marry Presbyterians or, Lord forbid, Methodists, specifically, Primitive Methodists. Then there is the matter of in utero damage to the foetus caused by crop dusting."

One afternoon, after a series of interviews in the seaside community of Raglan, Gillian wanted to hold interviews with people who had lived in Tokoroa when my family lived there .".. who may have first- or second-hand recall of your miraculous recovery, or recoveries to be more accurate."

She believed some of the attending behaviours of the Māori and Islanders may have "profound significance for initially attending to those who have experienced various forms of non- organic brain injury traumas."

She convinced me it would not be appropriate for me to participate in these interviews, but she was fully open to discussing who would be interviewed and where the interviews would take place. As a result, two interviews were set up, one with the second wife of Reverend Anderson, who, at this time, lived in Te Aroha.

While visiting there, another woman who had lived in Tokoroa also sat in. While I waited in the church with, by this time, the ancient Reverend Anderson, Gillian interviewed the two women.

Reverend Anderson showed me the oldest pedal organ in the Southern Hemisphere, and I asked if I could play it.

"I understand the conditional way in which you have asked," he replied, "and I have to say I wouldn't know whether you could or could not play, but if you are asking my consent, it would be appropriate to ask, 'May I play the organ?'"

"May I play it?"

"You most certainly may, Jeremy, but play something modern,

innovative, ecumenical, something North American that may challenge my ways of hearing God's music."

"I know the perfect thing, and I'll sing too."

Recalling the spirited repertoire of Kathleen Long, my Crofton Sunday School teacher, I began playing the American Episcopalian children's hymn, 'I Sing a Song of the Saints of God,' by Lesbia Scott.

*I'll sing you a song of the saints of God Patient and brave and true*

*Who fought and loved and lived and died For the Lord they loved and knew.*

*And one was a doctor, and one was a queen, And one was a shepherdess on the green.*

*They were all the saints of God, and I mean God help me to be one too.*

*You can meet them in schools, or in lanes, or at sea, Or in church, or in trains, while in shops,*

*While drinking tea.*

*For the saints of God are just folk like you or me. And I mean to be one too.*

While entertaining, engaging and enchanting my parents' old friend and former rector, Gillian engaged Reverend Anderson's second wife. Later, I listened to the tape recording of her preface to the interview.

"…The autonomic response to the lemon juice may have

triggered some brain function sufficient to allow the auditory functions to respond to the messaging of the singing and the direct requests of the Elders, 'you must come back.'"

Mrs. Anderson turned the teapot three times, a Kiwi custom. When they all had tea in front of them, she began.

"The boy was very thin, very weak, wan, but if you looked at him, you knew from his bright eyes he was going to live. He may not have known, but Rangi knew how to fix him. Rangi knew all kinds of things. Rangi knew people. She guided Eleanor Roosevelt around Rotorua. We usually add her family name twice to show we know she married her cousin.

Rangi would pick up the little Pakeha and carry him close to the geysers, even our greatest geyser, Pohutu, and give him nosey caresses down the side of his face, whispering Māori words to him. Sometimes she would sing him songs. Just to him."

The other woman added, "After the comas, Rangi carried him around on tours astride her hip, and he just hung there like a little jungle sloth."

Gillian laughed, and the woman continued.

"When he was tired, Rangi woke him as she roared to the tourists, 'Guests! Move along now! MOVE. Geyser is coming, MOVE NOW. Tourists! Move when Guide Rangi tells you to move. MOVE. Hurry now!"

Mrs. Anderson felt she had to provide some context.

"She got the people out of the way, and when they were all safe, she pushed her nose up to the little Pakeha's nose and stared him in the eyes. She taught him when he left her to shuffle one foot, like he didn't want to leave, then to turn his head, wink one eye, and stick out his tongue, which the Māori believe is the only part of the body a male can hide, then stamp his foot."

The other woman continued the story.

"Aye, that's how it was. If Jeremy didn't follow her directions, and he was just small, she would say sternly, 'Not so fast!' She mimicked Guide Rangi. 'I know where you have been. You know where you have been. You stamp your foot HARD on the ground. You wink your eyes TIGHT then open them WIDE. WIDER. You say goodbye like I showed you.

Old Hine recalled that they took the boy back to Canada, but at the Tokoroa Market, some people heard that the mother had written postcards to Amelia DeThierry's grandmother. If the boy heard anyone had been in Rotorua, he'd always ask, 'Do you know Rangi? She knows all about a Seventy Wind journey and the Sleep of a Thousand Dreams.'

Rangi used to say that people told her they had met a little Pakeha, and he'd asked them to tell her that Thor, the raftsman, had carried him on his shoulders, and that Sir Edmund Hillary met him up Huntly way, when they were heading back to Canada, after her Majesty the Queen made him a Knight of the Garter.

Rangi was to tell Old Hine from Tokoroa to look out for his old uncle if she ever passed through Christchurch, on her way to the Canterbury Plains. He'd give her a few pound notes. Imagine getting messages from a little Pakeha, part Canadian, part Scots-Irish cross.

Rangi knew people, looked them in the eye, and they never forgot her. Old Hine also never forgot. In the market, when new Pakehas arrived and people talked, Old Hine remembered the little Pakeha, whom she called her Canadian-Irish cross.

Everyone laughed and moved their toes in the gritty pumice earth. It was like talking about someone's pretty dog who got bought or sold, or stolen away."

# CHAPTER 3

**Dodge - 1986**

Although the town on the northwestern fringe of the Great Plains had another name, most people simply referred to it as Dodge. The original inhabitants of the area were Northern Cree and Beaver people. In the late 1880s, many were coerced into "selling their script" in Alberta, a deception that stripped them of land and autonomy. In the aftermath, they came to be identified by government decree as the Kelly Lake Métis.

The term "Métis," however, seemed to catch in the throat of the local child-welfare supervisor. He never used it. Perhaps pronouncing a word with an accent aigu felt like a betrayal of his loyalty to a rigid white-settler English. Or maybe the silence itself was deliberate, a quiet attempt to erase what made them different.

Among the residents were also descendants of the Iroquois, who had once traveled west with Alexander Mackenzie during his push to the Pacific. They had abandoned the expedition before Mackenzie reached the Nuxalk at Bella Coola. Some said it was a matter of survival. Others called it defiance.

In Dodge, the history ran deep, though rarely spoken aloud. The land remembered what the people tried to forget.

Although one white child, a girl, was born in the Peace River Block in 1908, the land wasn't made available for settlement until 1912, and the original farmer settlers and their descendants were referred to respectfully as "The Pioneers."

There were also Sudeten Germans who came to the area immediately prior to World War II, and they distinguished themselves as artists, artisans, teachers, and farmers.

They weren't allowed in Dodge or the adjoining community of Pouce Coupe until 1944. Like other settlers, they farmed. Others who had arrived before them had become successful grain farmers.

American soldiers came to build the Alaska-Canada Highway after Pearl Harbor was bombed. Later, some of these veterans and their families snuck across the border in the late 1940s and farmed the bottomland around places with names like Arras, Progress, and Hillbilly Junction.

During the Vietnam War, many extended family members arrived in town to escape the draft. Not as historically savvy as the Sudeten Germans, who were intellectuals and Social Democrats and constructed a Cooperative Commonwealth Federation (CCF) Hall in 1949, the draft dodgers established an annual Bluegrass Festival.

They welcomed others, men and women who resisted the Vietnam War. They and others formed back-to-the-land cooperatives that lasted a few years. Some were teachers, others came with theatre skills.

Some brought members of their extended families, including old women who were excellent quilters. Several of the men played the banjo well.

And there were the rednecks, described as "them's what lost the farm." Some of these people and their sons had found employment as "rig pigs" and made more money than they'd ever seen by working in the expanding oil patch.

Occasionally, I encountered them in the context of child protection investigations or child custody and maintenance applications. If they had a phone, and I called to schedule an appointment in my office, members of all the respective communities were quite loquacious, in a form of speech they called "rattling 'round like stones in a can." When I attempted to confirm the time and location of an interview, the likely response was, "God willing and the creeks don't rise."

Once, at a derelict shack on a semi-abandoned farm, a handsome, six-foot man, every inch severely impaired by moonshine, insisted that I, along with the police officer accompanying me, listen to him play "Silver Dagger." When I

attempted to confirm whether he could identify any kinfolk likely to be sober, he lifted his black T-shirt, revealing a hirsute abdomen.

Using a finger to circle the areola around one of his nipples, he smiled, then pointed to one of his three dirty toddlers.

"Who do you think y'ar, Pharoah's daughter? Git a rag and snot that kid!"

Attempting some degree of ecumenical empathy, or social solidarity, I simply asked, "Which one of them is Moses?"

\*\*\*

Whenever I had a free evening, I took Carole-Anne's advice and wrote about my childhood. The project was surprisingly helpful, if the measure of helpfulness was feeling I had a strategy to maintain some perspective other than staying under the thumb of a micromanaging supervisor.

After returning home from New Zealand, we moved to Crofton, where another pulp mill was being built. My paternal grandmother, the first teacher in town, had known of a piece of land with the best view of the Bay since 1904. In the *Quw'utsun Hul'Q'umin'um* dialect, it was *Sthixum*, or "little spring of water place."

With help from the Veterans' Land Act (VLA) and Jack Filberg, a family friend, my parents bought the piece of land my grandmother favored and "the little house" that came with it, on nine-tenths of an acre.

When we lived in the little house, an old tramp came around from time to time to ask my mother for a handout.

"Only if you have a song on your lips."

The tramp would grumble, and Mom would say, "And there's some wood that needs to be chopped and piled in the woodshed. Do you have a song?"

"You want me to chop wood and sing too? God, for the life of me, I can't remember words, and if I know some words, I mess up on the tune. I'll chop wood and stack it for you, Ma'am, but I don't think you can pull a song out of me."

Mom and I showed him the chopping block, axe, adze, ledgehammer, and hatchet and told him he could use the outhouse on the side of the woodshed. While he worked, Mom fried potatoes in bacon fat, cut off a couple of strips of bacon, then fried two eggs.

She said we should teach him that Harry McLintock song, "Hallelujah, I'm a bum!/ Hallelujah, bum again/Hallelujah, give us a handout to revive us again."

I went to ask the hobo how he liked his eggs and caught him by surprise as he swung the axe.

"Eggs! Eggs!" He let out a whoop, did a little dance, and started singing "Hallelujah, I'm a bum!"

I was thunderstruck and told him Mom and I wondered if we needed to teach him that song.

"Every tramp knows that song, sonny."

He sighed as he strained to put a heavy block on the chopping stump.

"I thought your mother wanted to hear a Wobblie song."

He went back to chopping and, when the axe struck, he sectioned off the block with the adze and the sledgehammer.

"What tribe are the Wobblies? They've gone extinct?"

He laughed. I wondered why everyone laughed when I started talking about extinct. *Extinct* was serious. It meant totally dying out. People, animals, all kinds of sea life and birds were wiped out.

"There are Wobblies who would laugh their heads off being called a tribe. No, the Wobblies ain't a tribe, they're bigger. And they're not extinct. Lots of Wobblies around Cumberland."

"I have an aunt, an uncle, and two cousins in Cumberland."

"See what I told you, Wobblies ain't extinct. You probably got Wobs in your fam-damily! Cumberland is a Wobblie town. Cumberland was where the Wobblies got their name. You had the Industrial Workers of the World, also known as the IWW.

So, this storekeeper in Cumberland tries to say 'IWW,' and he couldn't get it out; what comes out is 'Wobblie.' And because of that, Wobblies always joke around and call you a nickname. Some of their nicknames stick. Like me, I've got a friend called Slick- Eye."

My mother called me to take out a basin of hot water, soap, and a hand towel.

When I got back, the hobo said, "You again. I just started up again. Again and again, this child disturbs me."

"Mom says you have to wash before lunch."

"You go off and tell her not to put herself out too much. I'm enjoying this chopping. I'm also trying to work off a little winter blubber."

I repeated this to Mom as she passed me camp chairs for our picnic with the hobo. She also got out the oldest, cheapest knife and fork she could find, wrapped them in a serviette, and then got the worst set of mismatched dishes.

"Here, take these plates outside."

She told me to wait while she found another serviette, this one stained with blackberry juice or something. We had tried to wash it out, but the stain, now a blotch, was still there.

"Tell him not to wipe his nose with this."

Considering the food made up for Mom's silliness with the old cutlery and the serviettes, I said, "I'll tell him it used to be our favourite serviette and to look at the blotch and make up a story."

"You will do nothing of the kind," she said. "You can tell him he can do whatever he likes with this, wipe his hands or use it as a bib to keep his shirt clean. Above all, show him courtesy, which is the foundation of hospitality. Remember, courtesy, hospitality."

Mom had made coffee and put it into a thermos. The plate, featuring bacon, eggs, and fried potatoes, was set on an old, brown oval tray. On a pretty Chinese saucer, there were parsley and celery stalks filled with two kinds of cheese for me. We had gotten cheddar and white cheese when Dorothy Cameron, wife of Colin Cameron, the Cooperative Commonwealth Federation (CCF) Member of Parliament, came to welcome us to town. She had made the white cheese herself.

As we took the things out, the hobo had already taken off his shirt and was splashing water all over his chest and down his arms. When he saw us coming, he turned and looked over his shoulder from behind the towel he used to dry his back.

"Just practicing my after-dinner song, Ma'am."

"And the towel is your stage curtain?"

"Just getting my shirt on, Ma'am. I was looking forward to maybe cutting some kindling before you got out with the lunch."

"Well, I've got work to do in the house. If you need anything, you know who to ask."

I never eat eggs. I hadn't eaten them since the last time I was in a coma, and I sat back from the hobo so I wouldn't smell them.

Sometimes, even the smell of eggs cooking makes me vomit. "Your mother told you to sit way over there?"

"No," I replied. "I just don't like the smell of eggs. I puke." I started eating a piece of celery and a slice of cheese.

"Hey," he said. "Hey again! Doesn't anyone in this town say grace?"

I stopped chewing my mouthful of celery and cheese. "Oh, I know a grace."

"You think I don't?" I didn't say anything.

"Well, you gotta think and you gotta thank. And a kid shouldn't need to think when he's eating, but it sure goes down better if you pause for a moment before you dig in and give thanks. So what I had in mind before we dig into this feast your mom has put before us, is we bow our heads and give our thanks."

I bowed my head.

"If you close your eyes, you'll miss it," he said. "Bless these sinners as we eat our dinners. Amen."

"I was going to say, 'For what we are about,"

"Well, I beat you to it."

"I was also going to say that my uncle in Cumberland is named Bough."

"Really?" And what kind of a name is Bough? Wait. Don't tell me. I know. Wow. Bough. Bow-wow. Bough is a Wobblie name. Lots of Wobs up Cumberland way and everywhere have a nickname. Ya. For sure. Bough is a Wobblie name. And you can take that to the bank! That is, if you have any use for banks."

"Do you have a Wobblie name?"

"Sure! Lots of them."

"What do they call you?"

"Ask me no questions, I'll tell you no lies."

"No. Tell me your Wobblie name."

"Tell me yours, kid."

"I don't have one."

"Well, you better figure one out soon."

"I don't think I can. Why don't you tell me yours?"

"Alias."

"I beg your pardon?"

"Alias. Alias. Alias."

"You mean Mr. Alias?"

"Mr. Alias is fine with me. Go tell your mother this is darn good tucker."

"No. She told me to sit here and talk to you, showing you courtesy, which is a foundation of hospitality. You can tell her yourself when she comes to get the plates. Do you see the stain on the serviette? This used to be our favourite serviette, but then Nanna wiped loganberry tart off my face, and the juice stained it. Now it's a blotch. Would you like to look at the blotch and tell me if you can think up a story about it?"

"Where'd you get all this about blotches? You ever see a box of papers with blotches on them and somebody asking you what you see?'

"Oh sure. Rorschach tests. Ink-blot tests. Yes. The doctor I saw in Tokoroa showed me those after I came out of the Sleep of a Thousand Dreams. I'd seen them before when I came out of the Sleep of a Seventy Wind Journey. I liked looking at them."

"I didn't."

"Why not?"

"Oh, kid, I guess it was because of the person who was showing them to me. He was in a hurry."

"I think my mom wants you to sing a song."

"Your folks don't have a radio?"

"We have a radio, but we don't turn it on much now since Nanna got killed in a car accident in Ontario."

"Sometimes we get bad news."

"We listen to the news every night. We just don't sing along with the songs on the radio anymore."

"Well, I guess your mother needs a song. Or a poem."

"Oh, good!"

When Mom came out, wiping her hands on her apron, I told her Mr. Alias was going to sing her a song and recite a poem. She laughed and said, "Mr. Elias doesn't have to sing if he doesn't want to."

"Oh, he wants to, Mom, he really does."

"I think I'll just cut a few more pieces of wood, then pile  them, Ma'am."

"Oh, you don't need to do that."

"I've just got started, Ma'am."

"No, I mean you don't have to pile them. I'm sure I can find someone for the task."

"Would you like to hear a song then?"

"I'll never say no to a song, but you can keep the poem to yourself. Doggerel, I do *not* like, unless it's Robert Service. Do you know any Robert Service's poetry?"

"There are strange things done in the midnight sun. I know a poem for you."

"My father lived for a year in Paris with Robert Service."

"Well, settle up, Ma'am, and I'll tell you and your boy here about 'the men who moil for gold.'"

We sat spellbound and listened to the tale of "that night on the marge of Lake LaBarge."

When he finished, we both clapped, and Mom told him he didn't have to worry about the rest of the wood. When he started to pick up the axe again, she said, "That's not necessary. You might want to take the hatchet to make some kindling, and I'll fill a jar of coffee for you and wrap it up to keep it hot."

Then she asked me to come into the house, and I protested. I wanted to be outside with Mr. Alias, but she insisted that he needed to tidy up and leave.

He must have gotten the point because he chopped a bit of kindling and then knocked on the door. Mom and I were still doing the dishes when we heard the faint tapping and singing.

"So long, it's been good to know you."

We went to the porch door, surprised he was leaving so soon. "Well, here's your kindling, Ma'am. I better be movin'

along. And thank you for the fine meal and the good company."

"Do you know 'Hallelujah, I'm a Bum?'" Mom asked.

"Oh, I sure do, Ma'am. If that's what you want me to sing, I'll be singing it as I move along."

"Well, good luck to you, Mr... uh... what did you say your name was?'

"Elias. Elias, Ma'am." He started singing, "Hallelujah, I'm a Bum!" and I joined in.

My mother grabbed my arm and said, "Let Mr. Elias sing, we may learn a new verse or two."

He started to walk away, and she said, "Oh, wait, I have some things for you."

She left us standing by the porch and hurried into the house to get the things she'd fixed for him.

"Well, have you figured out your Wobblie name yet?" he asked.

"I haven't even thought about it, but I guess I should have one."

"Well, if you don't want to get called Kindling, you better figure one out."

"I don't want to get called Kindling."

"Well, you better think of one or that's what you'll end up with. Here's that special serviette of yours."

When Mom came out with the bundle of food and the jar of coffee, Mr. Elias packed them away in his duffel bag and carefully slung it over his shoulder, like he wasn't in a hurry to say goodbye.

"Well, thanks again, Ma'am, and it was sure nice to meet you." He paused, and I thought he was going to call me Kindling, but he called me "Fella," instead.

He didn't say anything as he walked to the road, and when he was halfway there, he started to sing "Hallelujah, Hallelujah, Hallelujah" to a tune that sure wasn't Hallelujah, I'm a Bum!

# CHAPTER 4

**Dodge -1987**

Dorothea, the local historian, mistook a United Church teach-in on Sexuality and Orientation for Sexuality and Ordination, or so she said. Once there, sitting in a circle among a group composed mostly of older women, she stated she always found Mildred "a comely woman." Then she asked if anyone would think it untoward if Mildred, an eighty-year-old Social Gospel feminist, came to live at her house.

Everyone except two lesbian farm women who taught agriculture at the college, and I, said it would be fine.

"What's there to think?" someone asked.

Glancing at Yvonne and Suzanne, the college instructors, I got a smile from Yvonne and a nod from Suzanne. I took this to mean "go for it" and said, "I don't know, Dorothea."

"Well, that is that!" she said.

She had raised her voice beyond what would have been an appropriate level and tone.

"I will *not* invite Mildred to move into my house. I have always been a respectable woman, a married woman, a mother, and the widow of a judge. I've enjoyed the fruits of marriage and have always had my own point of view, and if others disagreed, I was always intellectually honest. There have never *ever* been allegations of any form of scandal, but if one person, even a newcomer, would speculate that I, a good widow, was not a decent lady, I will not set myself up for scandal."

No one responded. Perhaps we should have, and that may have stopped the exhortation right there.

Dorothea went on. "I know how people can talk. I know what they would say. It would not be pleasant, and it would worsen when the woolie-headed, and there are more and more of them every year, would say it is my own business. Fie. Fie. I say, fie. I couldn't hold my head up in town."

She surveyed the room, pausing to glare at me.

"No! I will *not* pack my bags to leave this town in shame. Shame!"

Again, she glared at me.

"I've got my answer. I will *not* ask Mildred Moffat to move into my house, and that, ladies and . . ." she paused and adopted a sarcastic tone, "gentleman," she nodded at me, "is *that*."

Everyone present sat in silence, then she added, "I will say no more and trust to God no one will disclose what manner of discourse was raised here tonight."

We sat in a stunned, uncomfortable silence, avoiding eye contact with one another. What could anyone say to counter her outrage, her oratory, or God forbid, spark another diatribe?

Finally, Mildred spoke.

"No one asked my opinion. I don't even like her house. I like my own house perfectly well, thank you very much. I hope no one takes offence. She's done this kind of thing for years. Dorothea, just sit here and keep quiet for a while. It's time you learned a thing or maybe two."

## Eight Bells - 1956

No sooner had we entered the house after seeing off Mr. Elias than we heard bellowing.

We ran outside where, at the end of the driveway, we saw Mr. Elias crouched by the side of the road by Eight Bells' driveway. He clutched his left knee, and the contents of his duffle bag were scattered all over the side of the road.

"I've broke my leg! I've broke my leg!"

We ran down the hill toward him as Eight Bells came down his driveway. He knelt by Mr. Elias, put his hands on his leg, and felt through his pants.

"Just breathe deeply. Keep on breathing deeply. I think you've dislocated your knee-cap."

He cupped his hands around Mr. Elias's knee, twisted them slightly, then removed his neck scarf and wrapped it above the knee. He turned to my mother.

"Pat, could you get some ice?"

Mr. Elias moaned. "Now I don't know what I'll do. I won't be able to walk for a week or two. What will happen to me?"

Mom sent me back to our house to get ice, "or anything in the freezer you can carry." I hurried to the house and left her with Mr. Elias and Eight Bells. Inside, I grabbed a couple of ice trays and two or three slabs of frozen meat wrapped in brown butcher paper and carried them to where Mom stood. Eight Bells sat behind Mr. Elias with his legs placed on either side of him and his arms around his shoulders.

"He's in shock. I'm giving him the physical contact of my body. Keep on breathing deeply. You're in shock. Just let yourself feel me holding you, and the shock will pass. Now, Jeremy, pack the ice around his knee."

While I was gone, either Eight Bells or Mom had rolled up Mr. Elias's pant leg and bunched it above his knee. It was already swelling, so I placed two slabs of frozen meat next to it and then put the aluminum ice trays on top.

Eight Bells told Mom to go into his house for the first-aid kit she would find underneath the kitchen table.

"Bring me a triangular bandage. On the double."

I stayed with Eight Bells and Mr. Elias while Mom went to get the bandage. Eight Bells was still holding Mr. Elias.

"Now see how calm you are," he said. "You'll be right as rain in no time."

"I don't think so," Mr. Elias replied. "The last time this happened, I was laid up for three weeks."

"Well, you can stay here for three weeks," Eight Bells said, "but you must tell me a bit about yourself. Where have you just come from? Where are you heading?"

"I've just come from a stint in the Big House over a vagrancy charge, but it was all bogus. I was on my way to Salt Spring to visit old friends, then I was going to go off to Seattle to ship aboard a merchant marine vessel.

Mom returned with the bandage, and Eight Bells disengaged himself from the conversation. He started wrapping up Mr. Elias's knee. Mom said we'd better get the beef back in the freezer and be on our way.

Eight Bells interrupted. "It looks like I'm going to be having company for a week or so, and could sure use one of those packages of beef. If you go back to the house, there are a couple of rounds of Shropshire Cheese in the icebox. May I trade a round of cheese for one of the pieces of meat because, after it thaws, our patient here will need some good beef and broth to begin healing?"

Mom agreed and took me with her into Eight Bells' shack. We immediately were startled by a large parrot in a big birdcage who asked, "Who goes there? Who goes there?"

I introduced myself, and Mom and the parrot responded in a sing-song manner.

"They all be harlots, mate. It's off to Norfolk Island for yoooo; it's off to Norfolk Island for yoooo."

Although the shack itself was small, the yard had a shed with a raised pallet where Eight Bells did back rubs for men who'd strained themselves at work or had a bad back. I'd never been in the yard before, but it was something special. He had outdoor "rooms" hacked out of blackberry thickets. There were hanging baskets with various plants growing in pots, and five brightly colored cloths with printed tribal designs hung in the "doorways" to the "rooms."

There were a couple of old rattan chairs, and I plunked myself down while Mom went into the house. She came out quickly with a round of cheese wrapped in old, yellowed, ragged cheesecloth.

"Let's be on our way."

When we got down the road, Eight Bells asked Mom to help him lift Mr. Elias, and I was told to collect his things and bring them to the shack. Mom and Eight Bells supported him as he hobbled up the trail and helped him lie on the raised pallet where Eight Bells did his backrubs.

Hanging in the blackberry vines behind the pallet was the biggest tapa cloth I'd ever seen. Eight Bells asked Mom to put on the kettle for tea. As she lit the fire, I watched him gently place a couple of cushions under Mr. Elias's leg.

Mr. Elias was still in shock, mumbling that he was likely to miss his connections and would never make it to Seattle in time to ship out.

Eight Bells put one of the coloured sarong cloths over the parrot cage.

"By way of introduction, you can call me Eight Bells."

"The last watch," Mr. Elias said. "You're a sailor."

"A tar. I jumped ship in Penang in 1941 and have been here on and off since then, but you mustn't breathe a word if you ever get on your feet. If I hear, it'll be more than your bum knee you'll have to worry about. You're going to have to stay put until that wound heals, and that means if customers come around, you'll not be wanting to tell any of them you've recently been a guest of Her Majesty."

"Aye, aye, Captain."

"Oh, I was never a Captain. But I've got a safe harbour here, and I want to keep it that way. From time to time, people who come into port bring me things from around the world, and because I've got healing hands, whenever someone needs a backrub, they come here and pay me in beer. Everything else I grow, or trade, or people give me things."

Mom brought out the tea and said we should be on our way.

On the way back, she said we should be thankful Eight Bells was home or we'd have Mr. Elias on our hands. Then she told me that although Eight Bells told Mr. Elias he had jumped ship, we were never to tell anyone, as it may cause problems for him. Also, it was suspected that Eight Bells received Irish Sweepstake Tickets from various sailors who visited him. If he were caught, he'd be put in jail since gambling with sweepstakes tickets was illegal.

A few weeks later, while walking back from the post office, no beer bottles were cooling in the ditch in front of Eight Bells' shack, so we figured he and Mr. Elias had gone somewhere.

I had visited them only a few times, but I missed them both.

There was no one quite like Eight Bells. He had long hair, worn tight to his head, and had a braided pigtail down his back. He had been across the equator several times and had little gold earrings in his ears, one for every time he'd crossed.

Once, while I was visiting, Mr. Elias asked Eight Bells why he didn't have a tattoo. Eight Bells said a tattoo or scar can identify a man, and, although he had had opportunities, he never got one. As he was always concerned with his health and that of his mates, he had done his best to avoid scars.

# CHAPTER 5

**The Sprezza-Touras**

The town we lived in had the deepest harbour on Vancouver Island, which is why it was where ships came to take the copper mined in the hills. Between the wharf and the beach, there were huge black slag piles left over from smelting the copper.

A brother-in-law of Lieutenant-Governor James Dunsmuir managed the operation, and he smelted and shipped out so much copper between 1902 and 1908, the world price was brought down. He and his wife then left town, leaving piles of slag that washed down to the beach and mingled with the shell midden where the Quw'utsun peoples harvested shellfish.

My brother and many older kids played in the slag piles, but I was never allowed to, as my mother declared it unsafe. However, I preferred staying at home anyway.

Once, one of the Sprezza boys found an old Spanish sword there. My grandparents were visiting from Courtenay, and my grandmother said it should be taken to the provincial museum in Victoria.

Mr. Sprezza took it to the pub and sold it. He received a lot of money for the sword and became so drunk, he went home and got into a fight with Mrs. Sprezza. Mr. Talbot, who lived across the street, came to our house and told my father that if he didn't try to break up the fight, there would likely be a murder.

Dad had been in the front garden when Hiram Talbot arrived.

My mother grabbed the phone to call the police, then bolted up the hill to fetch Bill Pacheluk, the fire chief. By the time Dad reached the Sprezza house, everyone called them that, though their name was some long, double-barrelled thing, he found chaos.

Mrs. Sprezza was clinging to the shelves above the kitchen sink, arms stretched tight and trembling. Her feet kicked wildly at the air as Mr. Sprezza tried to wrench her loose, dragging her down toward the wood stove already burning beneath them. The flames inside popped and cracked.

She shrieked, kicking at his face while trying to pull herself higher, as if she could somehow climb into the rafters and disappear.

Dad froze for a second, stunned. Then he charged in, gripping his shovel tightly. By the time Mr. Pacheluk arrived, Dad had Mr. Sprezza in a neck lock with the shovel handle, and together he and Mr. Pacheluk held him until the police came. Mrs. Sprezza escaped and hid in the crawl space under the Talbots' porch.

The next day, she left with some of the children, instructing Mrs. Talbot to tell everyone she'd gone back to Alberta, but she only went to Chemainus. Within a couple of days, she had come back. Some people in town said she was a "glutton for punishment," which I couldn't figure out since she was the skinniest person I'd ever seen.

One afternoon, Mrs. Sprezza put on a new plastic raincoat and hat and came over to ask my mother if she would bake a birthday cake for one of the Sprezza children. Mom invited her in and asked if she would like a cup of tea.

"Well, maybe if you're offering, you might bring out your camera and take a picture at the birthday party."

Mom quickly explained that she would bake the cake, but she wouldn't be going to the birthday party.

"Then would you mind taking a picture of me in my new  coat and hat before God knows what...!"

She emphasized "God knows what" in a high voice, closed her eyes, and shook her head. "God knows *what* will happen to it."

"What could happen to your coat?"

"That's it! Nothing, if we've got a picture of it. If we ain't got a picture, well, God knows what."

"Oh, certainly, Mrs. Sprezza-Toura, I'll take a picture, but you must stay and have some tea."

"Seein' as how you insist, I will."

Mom took a while getting the camera and loading a new roll of film, so I got the tea ready. While putting cups and saucers on the table, I asked Mrs. Sprezza, who was having the birthday.

"Everyone!"

"On the same day?"

"Sure. Why not? Saves plates."

She stared at me, shaking her head and talking to herself. "Some smartass people ain't got a pinch of coon shit worth of sense."

Mom came out of the bedroom with the box camera and an

armload of clothes.

"May I ask whether you'd like to put on a sweater, Mrs. Sprezza-Toura?"

"You can ask all you want. I may say yes, and I may say no."

"I've had the windows open," Mom explained, "because we were painting, and I wouldn't like to see you catch a cold."

My mouth dropped. Mom held the cashmere pullover her American nieces had sent her. Betty had worn it first, then Maureen wore it for a while, but left for Paris to study and sent it to Mom with a note.

"Just couldn't wear anything like this and live on the Left Bank. Do I want *bourgeoisie* written all over me?"

It was a beautiful sweater with slightly puffed sleeves, a ruffled collar of a different stitch, and small triangular leather buttons. Mom had worn it a few times, then said it was too tight for her. Dad said it looked fine, but she said she hadn't been having any luck getting slimmer.

I had tried the thing on a few times over my undershirt and again with my bare skin under the cashmere. It felt better than silk, though the colour could have been brighter and lighter. I never really liked dark beige or taupe, except for car coats, but to give it to Mrs. Sprezza? My mother was known for her generosity, but this was plain silly. Where would Mrs. Sprezza wear a cashmere sweater? She didn't attend PTA meetings, and the family didn't attend church or the ratepayers' association.

"Maybe my mom won't be able to make the cake, Mrs. Sprezza, you not knowing whose birthday it's going to be and all..."

"It's one of the boys, Mrs."

Looking at me, she snarled, "I know when my kids were born."

"It'll need to be ready by Wednesday afternoon," she said to my mother, "next week, after school, and I can pay you in apples. But I would sure appreciate it if you could come down and take a picture of the party. It's a keepsake for the kids to remind them that we've had a happy life. Child Welfare would like that too. Good PR."

"Why don't I lend you the camera, and you can take the picture yourself?"

"I've never taken a picture, and the old man said if anyone ever brought a camera into the house, he'd be out the door."

"Well, Mrs. Sprezza," I interjected, "take the camera home with you and put it right on the kitchen table under the lightbulb where he's bound to see it."

Mom gave me one of her censorious looks.

Mrs. Sprezza said, "We'd better talk about the cake so as to get it right."

Mom agreed. They talked for a while and settled on a two-layer, rectangular applesauce and chocolate cake with icing. "Happy Birthday" would be written on it in blue icing, along with eight question marks and one exclamation mark.

She told Mrs. Sprezza to go into the washroom to try on the sweater. I could have died then and there.

She came out wearing it, and when Mom asked her what she thought, she said, "It'll do."

Then she was on to the Irish linen serviette. With a stubby pencil, she had drawn a big rectangle and scrolled around the edges a couple of times, pressing hard. The marks looked like they'd never wash out. I guess she was trying to show that she wanted a two-layer cake.

"Don't you think the sweater suited Mrs. Sprezza?"

"She looks positively bourgeoisie, Mom, positively bourgeoisie."

"That's the silliest thing you've ever said. The poor lumpen woman couldn't even look petite bourgeois."

"She's not even lumpy."

"Lumpen. A lumpen proletariat."

By now, I was crying, not only because of the loss of the sweater but also because my mother had criticized me. I felt so angry I used words I didn't understand. I will never know why she gave such a beautiful thing to Mrs. Sprezza.

"She's not even *lumpen*!" I said as I stormed off to my room for a good cry.

"You're probably right," my mother replied, "maybe one generation removed from being a landless peasant, but she asked for a cake and said she would pay me in apples. In a year, I'll send her off to a Women's Institute Convention. Just watch!"

Mrs. Sprezza didn't wear the sweater for long before Selina started wearing it all the time. Everywhere.

***

Depending on whom one spoke with, the birthday party was either the Pee Wee Party or the Razzmatazz Shindig. I took the cake down to the Sprezzas' house in my wagon and was told to come straight home after delivering it. I could go back later to take the picture.

When I got there, Mrs. Sprezza and some of the younger children were running around the kitchen, and when she hit the wash pan with a handful of spoons, the children started jumping up and down. She invited me to help.

"Don't just stand gawking, help with the cleanup!"

I couldn't see any cleanup getting done, but one of the children was out of breath. This probably was how they got the dust down under the floorboards.

Mrs. Sprezza had hidden hot dog buns and a jar of mustard under her raincoat on the table. When she and the children weren't jumping up and down, they lifted the coat and pulled a chunk out of a bun.

They all thought this was funny and said they would tell the bigger children who came to the party that rats had gotten at the food. However, there was still plenty of good eating if they watched where they bit.

Even Mrs. Sprezza took pinches out of the hot dog buns. "There will be food left over after the party," she said by way of explanation.

I wanted to go, but I made it as far as the yard. As I took the cake out of the wagon, Mrs. Sprezza yelled at me.

"What the hell are you thinking you're doing?"

"I'm bringing the cake inside to put it by the hot dog buns."

"You want the rats to get at that, too? Spoil the good cake your mom made for us?"

I didn't know how to answer. After all, Mrs. Sprezza was one of the rats herself. But I said I had to take the wagon home to get the camera.

She said I could get the camera, but not with the wagon, and if I wanted to stay around, I could help start the fire. I felt trapped and hadn't known how weird the Sprezzas were until Mrs. Sprezza came out of the house with an armload of clothes.

"What are they for?"

"Well, they're not Hallowe'en costumes. These are our clothes that people give us. Can't you see we're doin' a cleanup?"

"Are you going to get the kids to wash them?"

"You got rocks in your head? Some of them are for the fire, and some are for the bigger kids to make smoke signals."

All the Sprezzas smelled bad enough as it was, but I couldn't see how stinking up the clothes over the fire would make them smell any better. As I followed Mrs. Sprezza into the house, I asked what the smoke signals would say.

"Anything you want. 'Birthday Party at Sprezzas.' 'Smoked wieners.' Anything you want."

She brought out a stick from the stove and started burning the boxes the hot dogs came. Then she called one of the children to bring out the apple box.

"Get it yourself!"

Another asked, "What are you going to set on if you burn that?"

After a tongue-lashing that included, "stupid bloody bugger," Mrs. Sprezza asked, "Do I look like I'm going to burn the best chair in the house? What am I going to sit on while I'm tending the fire?"

As soon as she sat down after lighting some of the clothes, I'd had enough. When she wasn't looking, I took the cake out of the wagon, put it on the ground, and fled.

"Good riddance!" she called after me as I scrambled home.

"How do things look around the Sprezza house?" Mom asked.

"I really don't know where to start."

"Well, that's unusual. Did Mrs. Sprezza thank you for bringing the cake?"

"I think so, in her own way. They asked me if I wanted to have a pinch of the hot dog buns."

"A pinch of the hot dog buns? Did you sample them?"

"No. The kids said you had to be careful not to eat the parts the

59

rats had been at."

"Oh, you exaggerate."

I went to my room and looked at my pressed flowers until Mom called to ask if I was ready to go back and take the picture.

"The camera is ready. Walk. It might be a good idea not to take the wagon."

"Good idea."

As I headed down the road, I saw smoke coming from the Sprezzas' house. When I arrived, all the children who played at the slag piles were trying to cook hot dog wieners in the smoke. As soon as they saw me, they lined up and held onto their wieners like they were taking a pee.

I told them to "say cheese" when Mrs. Sprezza started yelling.

"You just hold your horses a damn minute! I want to get in the picture too."

She held a handful of wieners, and when some of the children yelled at her to "be a sport," she struck a pose like all the boys.

"Smoked wieners!" she hollered. "Get your smoked wieners!"

The kids repeated this until Mrs. Sprezza screamed, "Hurry up, you silly bugger! Do you think I can keep this crowd in line all day long?"

That night, Mr. Sprezza brought Mrs. Sprezza to our house with a bag of apples. He had been drinking, and he wanted to buy the film from Mom's camera. She graciously said there was no need to pay for the pictures and that she would have them developed and give them the negatives.

Mr. Sprezza said it wasn't a question of the money; he just didn't want any pictures taken, especially of his wife.

"Flashing her ass all over town!"

My father told him to go home and sober up. His voice was flat, quiet, but firm.

Meanwhile, Mom took the film from the camera. We'd only taken a couple of pictures, but even those felt important.

She handed the roll to Mrs. Sprezza, who held it delicately, as if it might contain proof of something no one would believe without seeing.

But before she could slip it into her pocket, her husband snatched it from her hands.

He fumbled with the canister, turning it clumsily until the film unspooled between his fingers. Light hit the surface. The images were ruined.

Mrs. Sprezza grabbed at it, desperate to save what was left.

He pulled it free like a belt and held it high, grinning at the destruction.

That was when she screamed.

She turned and ran, her cry echoing as she fled down the street toward home.

"He's got a whip! He's got a whip!"

Mr. Sprezza chased her, and when he reached the edge of our property, he turned and hollered, "Don't be giving her any more clothes to run away in!"

Both Mom and Dad were upset and talked about calling the police, but the Sprezzas fought all the time, and they decided against it.

Around this time, we bought a camping trailer and began camping on weekends with the Independent Order of Foresters.

Later, we became charter members of the Dogwood Trailer Club and visited various campsites throughout the Island. We met sane, successful people who enjoyed each other's company as campers.

When the school prepared for the annual Fairbridge Track Meet, someone gave Selina a blouse to wear while running. When the practice race ended, one of the teachers told her she would get the sweater back in a couple of days.

Someone was going to wash it for her and fluff it so that she could wear it when her picture was taken after winning the senior girls' race.

Selina went crazy. She wanted the sweater back immediately.

She assaulted a few of the mothers and wanted to look in their purses to see who had it.

When she realized she wasn't getting it back, she began to cry and ran every which way all over the field.

"Look at her run!"

"Like a high-strung thoroughbred!"

"That's fine, Alberta breeding you see. We don't sire anything quite like that in these parts."

"I hope she wins. It would do the girl such good to get a little respect."

"She's really such a good girl, despite it all. It's quite remarkable."

Selina walked back to where the women stood. "Nobody steals from a Sprezza! I'm not going to throw your goddamn race for you so you can stick it up your ass."

Some of the women were upset about Selina pushing them and trying to look inside their purses. Nor were they pleased with her accusation of stealing when they were only going to wash the sweater.

"Stealing!" Swearing! Did you hear what she said about throwing the race? No one said anything about throwing the race."

"Everyone has just encouraged her to compete."

Someone should have run after her to assure her they'd get the sweater back, or at least offer to accompany her to the woman who would be washing it and possibly stay for dinner. If someone had waved the sweater at her, she would have come back. She loved it so much, but nobody ran after her.

During the next few days, Selina repeatedly said she wouldn't be in the "damn race."

Miss Summerton, her teacher, said, "So be it. But I will ask you not to use profanities."

No one, least of all Miss Summerton, explained what profanities were. Compared to what Selina regularly saw and heard around her house, a "damn" or two probably held no real significance as a swear word.

What was obvious to everyone was that without the sweater, she appeared to have lost all her power and moped around like a whipped puppy. She only nodded when asked anything.

"Fine with me. As if I have any say in the matter. Any matter."

When she finally got the sweater back after school on a Friday, it looked as good as new and had a small envelope pinned to it in a plastic bag. The attached note read, "Zero sweater soap. Wash in <u>cold</u> water."

Selina took to wearing the sweater again all the time. She even ran in it. She also told people she slept in it so that no one could take it. It always looked wonderful.

People thought she was washing it with Zero, but she told me she wasn't washing it at all since she wasn't sweating, and she stayed away from the fire at home.

At school one day, she received Zero wrapped like a box of chocolates, and the card on it read, "Please race for us. We'll all root for you and pray out loud that you win. You are the best Sprezza we know."

The night before the track meet, she took the bread bowl to wash her sweater. The instructions on the box advised letting it soak, squeezing out the suds, then rinsing it. She lifted the bowl and fluffed the suds with a fork.

Her father, seeing the soapsuds, decided he would shave and take the rest of the box of Zero to the pub. Maybe he could get a couple of beers for it.

The men in town were nicer to him than usual, as some people had made bets on whether Selina would win the big race.

When the water was cold, he poured hot water from the kettle into the bread bowl, following which he and Selina got into a fight. She escaped to the bedroom she shared with her brothers and cried herself to sleep.

When she awoke in the morning, her mother had washed the sweater in the shaving water and dried it on a stick by the fire. As a result, it had shrunk, and there were pieces of Mr. Sprezza's whiskers sticking out all over it.

When we were meant to board the bus for Fairbridge, she stood by the door loudly crying and wiping tears from her face with her forearm.

Still, she got on the bus and made it to the start of the race. But when the starter pistol went off, she just stood there. By the time she finally started, the others were already far ahead.

On the bus ride home, she elbowed anyone who came near her, shouted disownment of her family, and that she wasn't going home.

When the bus stopped at the school, she got off and ran. If she had run as fast in the race, she would have won a silver trophy. No one ran after her.

She went home, if only because she had nowhere else to go.

The fight started almost immediately. Her father was the worst of it, shouting until his voice broke, fists pounding the table. But her mother was the one who jabbed her with a hot stick.

The police took her to the hospital.

There, Selina finally spoke. She told them how awful her family was, how her father used his shaving razor to slice bread because he didn't trust Mrs. Sprezza with a knife in the house.

After that, Selina went to live with people from the Salvation Army.

Miss Summerton returned to the house for the sweater. She brought it to Mrs. Whittaker-Ramirez, an Anglo-Irish lady from Chile, part musician, part scientist, and part mystery.

She experimented with the cashmere, trying to clean it without ruining its softness. In the end, the fabric came out darker than before, but intact.

Someone from the ladies' sewing circle took up a collection.

They bought a beautiful doll.

Eileen Kathleen O'Malley, Mrs. Whittaker-Ramirez's granddaughter, dressed the doll in a plaid skirt, a red cashmere tam, and the newly cleaned sweater.

The sweater had changed, yes. But it had been saved.

***

## Mrs. Whittaker-Ramirez and Eileen Kathleen

Her grandmother raised Eileen Kathleen O'Malley after tragedy claimed both of her parents.

Her mother, a beauty queen with a dazzling smile and a quiet sadness behind her eyes, died in a car accident not long after her husband, Eileen's father, was committed to an institution for veterans suffering from severe shell shock.

Mrs. Whittaker-Ramirez was not a typical guardian. Anglo-Irish by birth, she had arrived in Crofton on a Chilean freighter, bringing with her a cello, a set of copper etching tools, and a past full of strange chapters.

In her youth, during the wild decades of the 1920s and 1930s, she had lived in Montparnasse, the bohemian heart of Paris. It was there she met the love of her life: Ramirez, a revolutionary and a painter who smelled of turpentine and carried leaflets in his coat pockets.

She never spoke of how she lost him, only that she kept his name.

Sometime after World War II, Ramirez had been involved in political activities in Chile and was murdered. Mrs. Whittaker-Ramirez fled the country with her daughter, who, in turn, met and married a young soldier named O'Malley.

He was shipped off to Korea and returned severely traumatized from his time in a prisoner-of-war camp. It was rumoured throughout town that his wife was a "party girl who liked to bend her elbow." She was also known to enjoy "fast car rides."

When Eileen Kathleen was a toddler, her mother was killed in a car accident and, with O'Malley in an institution and there being no other family, Eileen Kathleen lived with her grandmother.

Since my birthday was in January, I wasn't allowed to attend school until I was almost seven years old. Instead, I spent a few afternoons each week with Mrs. Whittaker-Ramirez, who was knowledgeable in mathematics, chemistry, physics, and biology. She read to me about Marie Sklodowska-Curie when she should have been teaching me music.

In her kitchen window, there was an apparatus containing sixty-one beer bottles filled with various liquids, "thirty-six for the white 'keys,' twenty-one for the black ones" and a couple of small wooden hammers used to play Irish folk songs. The first song I learned was "Oh, the days of the Kerry dancing."

Eileen Kathleen didn't go to Guides. She stayed home with her grandmother and spent time making over her mother's clothes. At school, she always wore plaid skirts and blouses that fit perfectly.

Although she wore boys' Oxford shoes, she was by far the prettiest girl in town, and she liked to entertain.

Mrs. Whittaker-Ramirez was given the job of playing the piano for the school play as long as Eileen Kathleen could do a sailor's hornpipe dance in the middle of the show.

I visited them while they practiced. Eileen Kathleen played the piano while her grandmother danced on the table. My job was to hold a mirror so that Eileen Kathleen could watch her grandmother's feet.

# CHAPTER 6

**The Talbots**

The night before Mr. Elias and Eight Bells skipped town, they gave the parrot to Mr. and Mrs. Talbot. Eight Bells also left a framed painting of a four-masted ship, hand-painted on a piece of an old tea chest, along with instructions that it should be given to me.

Mrs. Talbot said the parrot, whose name was Scheherazade, had been on slave ships, but Mr. Talbot said it was easy to tell she had been aboard convict, not slave ships. When they raised their voices, Sheherazade shrieked, "It's off to Norfolk Island for *you- ooo*. It's off to Norfolk Island for *you-ooo*."

"There's your answer," Mr. Talbot said,

"Then we best start speaking in a civil tongue," Mrs. Talbot said, because sometimes Sheherazade can say things not fit for civilized beings or children to hear."

I asked if that meant Mr. and Mrs. Sprezza shouldn't listen, and everyone laughed. At the sound of laughter, Sheherazade said things like, "Not in this life, mite. Not in this life, mite. It's off to Norfolk Island for *you-ooo*."

Mr. Talbot explained that poor people from the streets of England, as well as the Irish, were transported to Australia for stealing loaves of bread. If they refused to work upon arrival, they were sent to Norfolk Island.

I asked why they stole bread.

"Many who were transported weren't thieves at all," Mrs. Talbot explained, "but upright men for whom there was no work in England."

Mr. Talbot added, "Some of them were sturdy beggars, some were rogues, and some were wild rogues, but they weren't bad men, just unfortunates on the nether ends of the class system."

I'd been to New Zealand and had met lots of Aussies who all seemed to be respectable enough as people. I asked if any women were able to go there.

"Just harlots," Mr. Talbot said.

"Mr. Talbot," his wife said reproachfully, "they were poor women and not all of them were women of easy virtue."

Sheherazade chirped in. "They all be harlots, mate. It's off to Norfolk Island for *you-ooo*."

Before I left, Mrs. Talbot brought us tea, and Mr. Talbot took down a large book from the shelf and said he wanted to read something to me.

"This is from a book entitled *The British Overseas, Exploits of a Nation of Shopkeepers,* by Mr. C.E. Carrington, M.A., from Cambridge University.

"From 1788 to 1866, 75,200 convicts had been sent to New South Wales, and 27,759 had been sent to Van Diemen's Land since 1817. At the time Victoria came to the Throne, approximately 1200 convicts were living on Norfolk Island, having been transported there for committing further crimes in New South Wales."

"They stole loaves of bread there, too?" I asked

"I presume they did something to irritate the authorities," Mr. Talbot replied.

When I went home, I asked my parents whether there were still harlots in Australia or had they "gone extinct." My grandparents were visiting from Courtenay, and my grandmother was horrified that a child would discuss such a subject.

***

## Waikato 1972, Why I'm Going to Mount Everest

The night Alice, Dawn and I *escaped* from the mental institution (we were non-unionized staff members), we were picked up by a jockey while playing kick-the-can-down-the-motorway at one o'clock in the morning.

He was the fastest speed freak on the racecourse, but the saddest jockey on the road, and when he asked me what I was working on, and I said, *Nirvana*, he asked why men climb Mount Everest.

"Once one starts climbing," I replied, "one forgets such trivialities."

Then he asked if, when I reached the top, I would have a hard time coming down. I told him I had already had a hard time going up.

His next question was whether I had thought of turning around and going back.

"I'm not alone," I said, "and besides, I've forgotten the way."

"What are you going to do when you get there?"

"There isn't much work up there. I'll probably do the same thing I'm doing here, only there'll be no distractions."

"What do you do now?"

I said nothing.

When he had no more questions to ask, I turned to Alice and smiled.

She was always the dreamer among us, with her pale hands clasped like she expected an answer from heaven at any moment. Alice wanted to see God face-to-face. That was her plan.

I couldn't help but smile again, this time at Dawn, who, of the three of us, seemed closest to her goal.

Seventeen, though she'd lied about her age to get the job, she lived like she was racing toward the end of something. Her dream wasn't heaven. She wanted ten thousand lovers.

She said it once, out loud and without shame, as if she'd already counted a few hundred on the way.

"I'm going to Mount Everest," I said softly, "because there's already one too many sad old jockeys in the world, and none of them understand why we left the mental institution at one o'clock in the morning to play kick-the-can down the motorway." We were talking "union."

\*\*\*

## The Flower Bee, Crofton, 1956

My first official contact with Canadian society occurred on July 1, 1957, during the 1957 Canada Day parade. The local rate payers association decided to promote the community as the Hawaii of the Islands. Those not involved with building the pulp mill became involved. For me, no fiesta nor carnival could have been as welcome.

We spent weeks decorating the float. The town's name was spelled out in bright crepe-paper letters, each one carefully pinned to chicken wire. That same wire held clusters of pastel tissue-paper carnations, crafted from toilet tissue.

We'd stack eight sheets at a time, fold them into narrow fans, tie the center with florist wire, and lift each layer until a bloom appeared.

I wasn't allowed at school then, so I had time, more than I wanted.

On my first afternoon at the community center, I made fifty-three carnations and three red hibiscus out of crepe paper. Someone asked how I'd learned to make them, and I said I'd spent time in tropical places.

"Ever seen a plumeria?" I asked.

No one had. Some didn't believe the flower existed until I drew it on the back of a flyer and described its scent, sweet and heavy, like rain and sugar and something just out of reach.

An old Italian woman, someone's grandmother, nodded slowly.

"Ah," she said, eyes lit with memory. "Frangio-pangio." With this authoritative pronouncement, other participants in the flower bee asked if I could get my parents' permission to attend the Hawaiian dance rehearsal that night on the public beach.

*** 

## Hawaiian Dance Lessons

A reel-to-reel tape recorder and several people with guitars helped a dance instructor from Saltair. Alongside a Kanaka-Black woman from Saltspring, the instructor taught Hawaiian dancing to anyone interested. I knew the Hak'ka and had seen a lot of Pacific-Islander dancing.

I spent the afternoon making flowers. Between the instructor's presentations, I ran in from the beach to the rhythm of a bongo drum and a few guitars, spontaneously bursting into an elaborate dance. I finished with a leap and a threatening pose with my tongue stuck out. I shouted, "Hungapie!" in as warlike a tone as I could summon.

Everyone was astounded and fell silent at first. Then, they burst into applause. A teacher, who lived in one of the cottages above the beach with his friend, the bank manager, arrived, carrying bongo drums.

"That was amazing! The kid is truly amazing."

Before the rehearsal ended, I secured a place on the roof of the truck that would pull the paper flower float. I told those already assembled about my crown, made of red, white, and black shells, and a Fijian grass skirt I could wear.

Within days, I had become familiar with everyone in town, and most people called me by my first name. The teacher with the bongo drums called me the "Kanaka Kid," a welcome change from "Davy Crockett." I had a raccoon-skin cap and a shirt with real plastic fringe, but I didn't identify with Davy Crockett, even though his ancestors came from the same town in County Antrim as my family, and we were probably related.

People dressed up as Hawaiians and marched behind the float, waving and laughing. The women on the float danced to ukulele music piped through a crackling speaker.

Some wore sari tops paired with fake grass skirts that rustled when they moved. A few wore black brassieres adorned with fresh flowers, tucked into their hair and strung through their leis.

I danced too.

I stuck out my tongue and winked at everyone along the route, like a little Māori warrior offering a fierce welcome. Most people smiled back. Some laughed and waved.

When we reached the end of the parade route, I stood on the float's edge and sang in a soft voice:

*Pō ata rā / Ka haere iho nei / E haere ana koe ki pāmamao (Now is the hour / When we must say goodbye / Soon you'll be sailing / Far across the sea.)*

The parade was a tremendous success. The wind picked up the final note and carried it somewhere beyond the crowd.

After the parade, everyone I knew in town went to great lengths to get me into Cubs and Sunday School.

# CHAPTER 7

**Wolf Cubs**

I forced my way into Cubs, much to the astonishment of Charlie Wolsey, the Cub Master. My brother, Jumbo Gentries, a neighbour boy, and several other boys were already in Cubs.

I wanted to *belong* to something. The school principal had described me as "categorically ineligible" to attend school and although my mother called the school board, nothing happened. She also called our MLA, Bob Strachan, Leader of Her Majesty's Loyal Opposition, who said the principal could make an exception.

The principal said he wasn't prepared to do so.

My mother called Aunt Barbara, who worked for the Filbergs, the richest family in the province. She drove down in Jack Filberg's Cadillac and said she could get me into Queen Margaret's School for Girls. My father had called from work that day, and Mom told him that Barbara would drive me to Queen Margaret's, but that it would cost us Nanna's silver-and-crystal cruet set.

"No way!"

The next thing we knew, he was home and parked behind the Cadillac.

"What are you doing home?" Aunt Barbara asked. "Family emergency."

Mom told me to go outside and play.

I went around to the back of the house so that I could listen. Dad explained that, in principle, he was opposed to children living away from their parents.

Queen Margaret's was a private school, and those who sent their girls there believed it would elevate them in the social hierarchy. Contributing to that would be against everything he had fought for in the war.

"Besides, it wouldn't be good for Jeremy," my dad said. "Half the time he already thinks, or at least wishes, he was a girl, and being around girls neglected by their own parents will put him in a situation where he would be vulnerable."

My aunt Barbara responded in an exasperated tone, "It's not as if he'd be a boarding student. He would be a day student, and only for one year."

"And who would drive him every day?" Dad asked. "Arrangements can be made, Fred."

"Barbara, we can't afford it. We can't go on living in this shack. I've got an application in for a VLA Loan for the house we're building."

"Mrs. Filberg is prepared to buy the cruet for whatever you're asking."

"Barbara, it's not mine to sell. It was Sarah Ellen's, and although I'm sure Mrs. Filberg would give us an honest price, it's not for sale. The sale likely would be enough to cover the cost of the tuition, but it's against my values to send Jeremy to any form of private school."

From where I watched, Aunt Barbara took a lace hankie from her pocket to wipe her eyes. Then she started to cry.

"I was only trying to help, Fred. Don't you realize Pat wants to go back to work? She ran the government office before and after you were married. She wants to work, and Jeremy being home means she can't work in government."

"Jeremy will survive, Barbara."

When Dad moved the car so that Barbara could back the Cadillac onto the road, I went into the house. My mother stood by the cutlery drawer, shaking it and rattling all the silverware. She was crying.

"I just want to go to work! There are absolute moron-men in every office in the municipality and the school board and, with the odd exception, in the provincial government. I just want to go to work!"

Wolf Cubs was a way of belonging to something official, maybe even having friends. I was prepared to trade or gamble anything to find out where the Cubs had their fort, and where my brother, the Gentries boy, and the other Cubs lit fires and had their secret meeting places.

At the meeting I tried to sneak into, Akela, as everyone called Charlie, was setting up his box camera on a tripod. I asked one of the Cubs what was going on.

"We're going to get our picture taken in our new uniforms."

"Great! I'm glad I wore my grandmother's fox fur stole."

Charlie began positioning everyone under the chestnut tree in the Anglican Church yard. Some of the mothers came to comb their sons' hair and ensure their knee socks were held in place by little green ribboned garters.

No one wanted me anywhere near the picture-taking. After a couple of winks and nods, one mother moved away from the porch and stood behind me with her arms crossed in front of me.

"It's the right thing to do," Charles said.

I certainly didn't agree. Turning around and staring up at her, I took the head of one of the foxes and shouted, "Madam! Madam?" until I got her attention.

"How are your teeth?" I asked. "I beg your pardon? "I was just wondering how your teeth are."

"I have no intention of releasing the hold I have on you if that's why you're going on about my teeth. I'm not going to let go of you, boy, until the rest of them have their picture taken."

I felt uncomfortable standing face to face, at least looking up into her face with this woman who held me against my will. I'd never met an adult as scary. She was the most serious adult I'd ever come across, and I resented her controlling me. I needed all my wits to get out of her control.

"About your teeth," I went on as she ignored me, "about a week ago, I put poison-dart oil on some of these fox teeth. I was going to pull them out and put them on blow darts, but I haven't got around to it yet. Are you in pretty good health? I'd like to try something on you..."

This woman hadn't been on the float and was new to town. What I didn't expect was her screaming and telling everyone she was pregnant.

"It wasn't me. I'm totally innocent!" I said.

All hell broke loose. Charlie couldn't take the picture. The women on the porch ran to her with congratulations and questions.

"When did you know? How many do you already have?" Charlie was furious.

"I'm designating you head boy," he said to my brother. "Now take your brother home."

I had no intention of leaving, but as my brother approached with his fists clenched, I thought it was wise to make a quick exit from the churchyard and cross the street to Tailor's store.

Scottie Drean, the only female stonemason on the Island, was chatting with the bank manager and his friend, the art teacher. They saw me running while angry and crying.

"What's wrong, Jeremy?" Scottie asked. I stopped and told them everything.

"This is total bullshit," the art teacher said. Scottie motioned. "Follow me."

The four of us marched across the street. A nasty squall was brewing on the bay, and from the mist in the air, we knew a storm was soon to come. I didn't think Charlie knew this, but maybe he did, which could have been why he was so anxious to take the picture.

Scottie approached the women, who, by this time, stood in a group on the lawn.

"It's supposed to be a picture of the boys whose families had bought uniforms," one of them said, trying to explain.

Another added, "It's been such a successful project." The rest agreed. "Aye, Aye. It has been good." Scottie would have none of it.

"My boy is in uniform. It will add so much to the picture if we have someone dressed like a field mascot."

Looking Charlie in the eyes, she strode toward him, and when she stood in front of him, they stared at each other for what felt like a long minute of silence.

She spoke to her son, Vaughn. "Tell this nincompoop he doesn't have my permission to photograph you if Jeremy, or any other child here, can't be in the picture!"

Vaughn Drean was the biggest Cub. Until then, he'd been watching the unfolding drama from the middle of the back row, where two rows of Cubs were lined up waiting to have the picture taken. Immediately on being summoned by his mother, he came to her side and laid his hands on his hips.

"Ya, Akela, if you want to keep your job, you've got to let Jeremy hold the slate."

"What's the big hairy deal anyway?" the art teacher asked

Charlie. "I've seen lots of Cub packs and Scout troops where half the kids didn't have uniforms."

"Not in *my* Wolf Cub pack you don't. Either the boys have a uniform, or they are not initiated as Wolf Cubs. And I don't like your talk. Don't pretend you don't know what I'm talking about. I don't like you saying, 'big hairy deal' in front of these boys. That's not proper and you know that."

"Well, henna my pubes," the art teacher replied.

Nobody knew what this meant, but Charlie was upset. "That will be enough! Enough! Do you hear me?"

The bank manager intervened. "All right, all right! When's your next initiation?"

"Next meeting," Charlie said.

"Well, why not initiate Jeremy right here on the spot so he can get into the picture?"

"I'd rather not."

"You may not have a choice, Charlie," the art teacher said, his voice rising over the growing noise. He waved to Scottie and gave the bank manager a discreet nudge.

Within minutes, a small group of women had gathered under the chestnut tree. We all leaned in, forming a loose circle, voices tumbling over one another in urgent whispers.

Charlie, meanwhile, was trying to organize the Cubs. He raised his hands and called out names, but the boys kept drifting like minnows in every direction.

Vaughn ran back and forth between Charlie's chaos and our huddle, breathless and grinning like he was born for this kind of mayhem.

"If we want to pull off a special tableau," Scottie said, tucking a loose curl behind her ear, "I can get some of the float dancers here in thirty minutes. Maybe less."

We all paused, half excited, half panicked. It might just work.

The bank manager pointed out that we didn't have time. Charlie turned his camera around, and Vaughn told us we were running out of time. He beckoned to me.

"C'mon, Jeremy, just as he goes to take it, we'll run through from opposite directions."

I told Scottie and the men that I would go for it, and I doubt even they expected to see what went on next.

I let out a war whoop and started running. Taking the scarf off my neck, I ran to Charlie and started whipping his behind. He ran toward me, attempting to snatch at the fox heads.

"Poisoned teeth! Poisoned claws!" I screamed. "Grab if you like, but it's you who dies!"

I ran back to the base of the tree and started to huck chestnuts. Hucking is any activity that involves picking up chestnuts (conkers) as fast as possible, then pelting (hucking) them at the target, in this case, Charlie's fat ass.

The art teacher helped me. Charlie tried to ignore us and take the picture. Vaughn and I, along with the art teacher, continued the chestnut blitz.

Jumbo Gentries insisted his eye was hit, but it was only his cheek. He bawled elephant tears, at which point a few Cubs tried to get my brother to come over to beat me up.

Then, someone appeared whom none of us were expecting.

Grandpa Chiko rode all over the Cowichan Valley on a beautifully tooled Mexican saddle with pieces of hammered silver over it.

His dog, Pale Face, and Diamond, his pinto, were well cared for, the three of them reminding us of a different time, yet there, under the chestnut tree, very much a part of the moment.

Everyone loved Grampa Chiko, and we all talked at once. He held up his hand and let Scottie explain that Charlie planned to take a picture but wouldn't allow all the kids to be in it.

"What some people don't realize," Charlie said, "is that this isn't a meeting of the ratepayers association."

We all laughed when Charlie, with his pompous manner, pronounced it "rate pears association."

While everyone tried to give their opinions to Grampa Chiko, Vaughn got under the black cloth at the back of the camera.

"Say cheese!" he shouted. The flash went off. Vaughn came out from under the black cloth and bowed to everyone assembled.

Scottie said to Charlie she would pay "for all the pictures if you make prints for everyone, and maybe a few for posterity," if Charlie brought copies to her house before the next meeting.

The mention of posterity sparked an eruption of pent-up laughter. By this time, Charlie was rubbing his backside, either from being pelted with chestnuts or maybe pricked by a fox tooth or two.

Lightning crackled across the sky.

Charlie told Scottie she could "whistle Dixie all you want, but you'll learn not to interfere with an artist."

Thunder rumbled and, within seconds, monsoon rain poured down.

Scottie, the best stonemason in town, perhaps on the whole Island had borrowed a truck to haul slabs of striking yellow-orange rock for her house on Arthur Street. She told the bank manager and the art teacher to get the truck.

"Jeremy, you get in too quick!"

Vaugh jumped into the back with the rocks.

We piled into the truck, Scottie behind the wheel, the art teacher beside her, and me sitting on the bank manager's lap. As soon as we closed the door, we started to laugh.

The harder it rained, the louder we laughed. As we jerked onto York Street, the bank manager hugged me and said, "Scottie, you're the hero of the day!"

"Jeremy Gabriel was the real hero," she replied.

"Vaughn has to take some of the credit," I said. "We all do," the art teacher added.

Scottie arranged for Eric, the harbour master, to act as a go-between to get the pictures, but Charlie said he wouldn't cooperate.

The art teacher and the bank manager moved to Vancouver, the former returning to university. He later became an art professor at the Nova Scotia College of Art and Design.

One summer, years after they had moved away and when I had my paper route, I rode past the beach cottage they visited on long weekends and over the summer holidays. The bank manager was in the yard washing a red sports car. He recognized me and asked me to come in.

He excitedly ran up the front stairs and into the cottage, calling for his partner, the art teacher. They both appeared happy to see me and offered me a glass of Coca-Cola with ice and a slice of lime.

Then they asked how I was doing, how I liked school, and if I had ever received a copy of the infamous picture.

I was happy to see them, but they could stay only for the weekend, as things were going well. They would come back and stay longer, but I never saw them again while I had the paper route.

They were true gentlemen.

They once sent me a postcard featuring a Greek god while they were on holiday in Europe. They wrote about their glorious time in Paris, suggesting that I should visit, as I would "especially like the statues." It was signed "Love," but I couldn't read their names, still, I knew it was from them.

# CHAPTER 8

**Legend of the Woman at the Mental Institution Bus Stop (Can be sung to the tune, Oh, the Days of the Kerry Dancing)**

Beside the Great South Road, near Auckland

In a dream or maybe nightmare

Sits a woman at a bus stop

She has sat there every day

For thirty years the legend says.

In a dream or maybe nightmare

With a sandwich lunch she sits

Feeding swallows when she's lonely

Blessing blossoms in the springtime.

In a dream or maybe nightmare

She's been waiting for her lover

Hands may shiver in the winter

Sweat may blind her in the summer.

Governments have come and crumbled

Miracles have been recorded

As she sits and prays to see

Her soldier lover is coming for her

## Oedipus and His Half-Wit Half-Siblings

I've forgotten the title we were officially meant to use for Captain Hennie. He would have been as happy had we shouted, "Scout Master Hennie, Second Class, Sir!" Since he had been a captain in the army during the war, he was referred to as "Captain" by fathers present at church parades, lunches with the fathers' group, or the more select committee meetings.

Mr. Hennie, the Scoutmaster, also worked shifts at the mill.

His eldest daughter had gone off to UBC, returning only for holidays, tall and elegant like a swan. Everyone said she had her mother's looks, or at least, the beauty her mother once had.

His son, for reasons no one discussed, never joined the Scouts.

The youngest, always smiling, had a mouth full of crooked teeth and probably should have had braces, but no one mentioned it, and she didn't seem to mind.

After the eldest daughter married the son of a senior manager at the mill, people began to speak of her with a kind of reverence, as though she'd fulfilled something for the whole family.

And when Mrs. Hennie came into a room, someone would always whisper, "She used to be just as beautiful," as if it explained something that couldn't be said aloud.

Mrs. Hennie never washed herself, cleaned her teeth, or brushed her hair. Nor did she clean the house or do laundry. She wore whatever would cover her, without concern for how it looked, whether it matched, was inside out, had been sprayed on by the male

90

cat, or had kittens born on it by the mother cat. No one visited her.

When she felt particularly sad or lonely, she came to our house, arriving at the back door, crouching down, and shielding her eyes as if looking through a keyhole.

"May I come in? Please, may I come in?"

Whenever she came, we never knew how long she had been out there, faintly calling while facing the doorknob. Eventually, as promptly as we could when we knew she was there, either my mother or I let her in. Sometimes, she was sad, cold, and visibly relieved to come in from the rain.

"Mrs. Hennie," my mother always began, "would you like a cup of good tea?"

"Oh, no! I just want to sit a while and visit." That was my cue.

"Would you mind if I watched TV, Mrs. Hennie? If it bothers you, I'll shut it off until the scary parts end. Or would you like me to say my lines for the play I'm in?"

Around this time, my mother would call me into the kitchen to help with the tea.

I knew the routine: I'd set out a cloth serviette, one of the pretty ones for Mrs. Hennie. Tea at our house always meant food.

My mother, among her many talents, was a remarkable cook, skilled in cuisines far beyond our little town.

Once I'd laid out the bone-china cup and saucer, and the matching bread-and-butter plate, she would emerge with the tea tray.

It held sandwiches, squares, or other delicate treats. There were often Nanaimo bars, perfectly cut and layered, just the way Mrs. Hennie liked them.

I watched the whole performance like a stage play we both knew by heart, my mother with her quiet elegance, and me as the assistant stagehand, passing her props.

She sat with Mrs. Hennie and asked if she'd heard from her daughter at UBC.

"No," was the inevitable answer. "You know how it is. I don't go down to check the mail, and when I do, I might not have my key.

Ivan doesn't like it when I go up to the wicket and say, "General delivery."

When she was excited, she spilled tea on her clothes, put the cup down, and stuffed food into her mouth.

"And Mrs. Gabriel, I haven't been working on my small projects. The fucking mother cat is pregnant again, and I think the tom cat is one of her babies from a litter last year. I don't know what I'm going to do, Mrs. Gabriel."

She'd laugh, slurp her tea, and, if she made a mess, she'd wipe her mouth with the sleeve of whatever she was wearing.

When she composed herself, she carried on. "I haven't been working on my small projects. The fucking mother cat, and now I'm going to have half-wit kittens shitting all over."

After one such encounter, I ran down the hall to my older brother Freddie's bedroom to tell him the Hennie cat was going to have a litter of half-wit kittens. I explained about the tomcat, and he looked up from his Periodic Table *of the Elements*.

"Why are you surprised?"

"Why shouldn't I be?"

"In this case, it's a matter of nature and nurture. How could they be anything but half-witted?"

"Are you trying to tell me that because someone has a mental illness, their pets are likely to behave the same way?"

"Just think about it. When you figure it out, come and tell me. Hint: think Darwin, or, better still, go tell Mrs. Hennie she could also call the tom cat Edie Puss, which is Greek for killing your father and sleeping with your mother."

***

**The Kennedys Are Not Yankees**

Although the rest of his life was profoundly disorganized, work, sending money to his daughter at university, and his involvement with the Scouts were meticulously organized aspects of Mr.

Hennie's life. He tried to be a reasonable, rational, and scientific person about everything he did, but he wasn't a very good organizer.

He took us into the bush frequently and taught us the names of trees, how to identify them, how to use a compass, mark a trail, and follow it. He showed us where an important Chief of the

Qu'amichans was buried. I got lost on that hike and was found on the Comaiken Reserve near the village of Clemclemalits, just before Mr. Hennie organized a search party.

One weekend in late November, Mr. Hennie took ten of us on an early winter camping trip.

We dug latrines and sectioned off the camp, as if we were establishing a base in hostile territory. I hated all of it.

I didn't mind lining up to see the Queen or marching in a parade, but digging a latrine wasn't my idea of a good time. I said as much, loudly.

That earned me KP duty.

My job was to keep the camp clean, the fire lit, and spirits high, apparently. If I failed, there would be "no evening hike before campfire," Hennie warned.

"It's all a matter of your attitude," he added, "betrayed by your insolent mouth, Gabriel."

I didn't answer. But I stoked that fire like it had insulted me first.

To a thirsty person, even a bitter fruit tastes sweet.[4] Soon, I had company and was damn glad of it. I knew the legendary Old Lady Holman, the best cougar hunter around, had shot a mother cougar around the camp a couple of weeks before, and a male cougar always comes looking for a mate.

I admit I felt scared. Any feline who had anything remotely to do with Hennie was probably disordered, or inbred, and might mistake me in my brown Scout uniform for its mother, or its mate, or something to eat. Either way, I was glad when Jake-Jack-John showed up.

Jake-Jack-John was known as the "baddest kid in town." Some pulp mill people who came from Ontario said his family had to move to BC because he had burned down a church.

But his father, an electrician, said, "It was faulty wiring.

Besides, the church was on Indian land, and they were never paid for it. They had no interest in the brand of holy rollerism preached there."

According to Jake-Jack-John, the Indians in Ontario had a bad enough time with the Catholics.

"Why in hell would they take up with anyone else?"

I got along well with Jake-Jack-John. He'd get me, or anyone he was with, into all kinds of mischief, but we'd usually get out of it. He was never violent with anyone else, nor was he a bully, and I liked the way he acted up. He had, for want of a better word, a talent. When he had an idea, his huge black eyes would sparkle like twinkling lights.

The weekend after President Kennedy was shot, Jake-Jack-John, Mr. Hennie, and I were on a surveillance hike around the camp. With everyone else ahead of us, Jake-Jack-John began talking about the assassination. Before he could say much, Mr. Hennie interrupted him.

"You don't begin a discussion about a tragedy by saying, 'The part I liked the most.'" Then, more sternly, he said, "There is nothing to like. Tragedy is a quality of experience that must merely be accepted."

Or avoided, I thought, but Jake-Jack-John continued undeterred. "Yeah. Yeah. Anyway, the part I liked *best* about the assassination, "

Mr. Hennie was now infuriated. "Do you not listen to a thing anyone says to you?"

"Listen!" Jake-Jack-John said, looking annoyed. "I'm only trying to express myself. And the *part* I'd like to express is..." He paused to see if Mr. Hennie had given up.

"When they interviewed the fat guy in the park beside where the motorcade went, you know, the grassy knoll. What I liked about him is that when he talked about how the president and Mrs. Kennedy's car was coming toward him, he started to blubber."

He thought this was hilarious and ran to an old cedar stump, trying to climb up as if he were one of the Three Stooges. Then he ran around Mr. Hennie repeating, "blubber, blubber, blubber" and "tree stooges."

Mr. Hennie chased after him, yelling as he ran.

"Get a grip on yourself, Dudley! Get a grip on yourself!" Jake-Jack-John rolled over the ground, kicking his feet in the air.

"They shot the president, blubber-blubber." He sat up and shouted, "They got me in the head. My pretty Yankee head."

Rising on his elbow, he said, "But they didn't get my pretty wife."

"You stupid boy! The Kennedys aren't Yankees! You don't understand a thing."

"Oh, the Kennedys aren't Yankees?! Now I've heard everything. They are so. *Are so, are so, are so!*"

"Are you calling me an arse hole, Dudley?

Jake-Jack-John rolled over onto his hands and knees and crawled around. Occasionally, he froze and cast a bug-eyed stare.

"If I had a tail, I'd be a pointer."

Mr. Hennie screamed at him again. "This is outrageous! Stop immediately! I order you! Stop this silliness immediately!"

Jake-Jack-John had pushed beyond what he could get away with before Mr. Hennie attempted to restrain him. He acted as though he were dying, like an actor in a Western, kicking his feet in the air, holding his heart, and making ghoulish choking noises.

I applauded, and Mr. Hennie yelled at me.

"Gabriel, you just encourage this insane behaviour. You're on KP duty. Again. Back to the camp."

By this time, Jake-Jack-John had caught his breath and pointed at Mr. Hennie.

"You just encourage! You just encourage!" When he got to "insane," he said, "Oh what the hell, crazy behaviour," hollering at the top of his voice.

Mr. Hennie was hollering too. Something about Yankees being a tribe from England and New England, *and the Kennedys are certainly not Yankees.*"

Now and then, when I'm in the woods by the Chemainus River,

I hear the wind rustling in the trees, and I can almost hear Jake-Jack-John yelling from far away. "The Kennedys are Yankees, and I know that for damn sure!"

Later that weekend, Jake-Jack-John and I and the rest of the Scouts got into a major battle with spears and sword ferns. I rode to the hospital in a police car with the siren blaring and the lights flashing.

Jack-Jack-John came along, and he told someone at the hospital, "That's his career shot. I mean, look at his mouth. He could have been an important film star. Now he's yesterday's news. I'd sue Hennie."

Shortly after this, Mr. Hennie quit as Scoutmaster. Everyone thought it was because of what happened to me, getting my teeth knocked out and all. It could have been for many other reasons.

When Jake-Jack-John returned from his confrontations, I was well into the rituals of KP duty and had lit a fire to burn camp refuse. Mr. Hennie had taught us to "burn-bash-and-bury" tin cans and whatever evidence of there having been a campsite. I also cleaned a big stew pot with a stick and sand from the river.

Jake-Jack-John said he would "just straighten up around the tents."

After a while, he called me. "Do you want to see something really amazing?"

He had been in Mr. Hennie's tent.

"I don't know whether we should," I said.

"Oh, come on. How can you say no? It's not as if I'm going to steal anything. I mean, what does Hennie have that I could want? Well, maybe there *is* something. C'mon in."

He had already gone through Mr. Hennie's kit. Various items were arranged all over his sleeping bag. Poking out from under the pillow was a magazine like none I'd ever seen. It had pictures of women showing their breasts and buttocks.

"I don't want to look at this," I said. "It's his private stuff."

"Private, my eye! One of them is a picture of his daughter. Look at the pretty one here. That's his daughter. Lookie-lookie."

I looked. "Jake-Jack-John, it is not."

"Yes, it is! Hennie's going to get his gun from the war and hunt down the photographer and shoot his nuts off. It's his daughter, that's why he's got it hidden under his pillow."

He folded his arms and summed up his discovery with a curious exclamation. "Investigation!"

I shook my head and felt it necessary to tell Jake-Jack-John to smarten up.

"This isn't true."

He stopped long enough to make a face.

"Yes, it is! This is happening. Hennie's going to shoot his daughter's photographer. Why else do you think he's got these pictures?'

"I think he likes looking at the pictures . . ."

"And?"

"He likes looking at them."

"Do you know why?"

"Because Mrs. Hennie has a mental illness and doesn't wash and all that."

"But I bet you he pokes her."

"I'm sure he doesn't."

"I don't know. Maybe he doesn't do it with anyone."

"He pokes himself. I've heard my father talking when Mr. Bank and Mr. and Mrs. Solly were over for cocktails. They all agreed.

Hennie is his own worst bumboy."

By this time, I was curious about the magazine. "Let me see that picture."

"Nope. It's mine."

"You said you weren't going to steal anything."

I reached for the magazine, but Jake-Jack-John pulled it toward his chest, then ripped the page out. I was horrified and speechless. It might be a magazine of naked women, but it was Mr. Hennie's property.

"Now you're in trouble! I said.

"He'll find out."

"Of course he'll find out!" Jake-Jack-John hissed.

"I'm going to make a WANTED poster with the India-ink pens. Then I'll nail it to a tree, so when Hennie comes back from the hike, he'll see his daughter's picture under the word WANTED."

He was giddy, flushed with something that felt like vengeance.

"And underneath," he continued, "in small print, I'll write: 'Captain Hennie wants the man who photographed this naked lady for bumholing… or ass-ass-ination of his daughter.'"

He reached for the ink bottle and twisted off the lid just as we heard the crunch of boots on leaves, Scouts returning from the hike.

The bottle slipped. Ink bloomed in wide, damning stains across his sleeping bag.

I froze. I hadn't finished KP duty. Jake-Jack-John had done nothing all day except rummage through Hennie's kit.

Then, with theatrical flair or total lunacy, he unzipped and started peeing on Mr. Hennie's pillow (and sleeping bag).

I scrambled to my feet, shoved aside the tent flap, and slipped out just as Hennie and the rest of the troop filed into camp.

I immediately shouted, "I didn't do it! I'm innocent!"

From inside the tent, Jake-Jack-John yelled, "I'm not coming out until I get my choice of weapons and my second to kill Hennie if I miss with my spear."

Mr. Hennie threw a tantrum.

"This is family mutiny!" He scrapped "family" then simply shouted, "Mutiny! Mutiny! After all I've done to include them!"

He grabbed me by the arm and bellowed, "Why were you in my tent?"

I knew mutiny was serious stuff. I'd read Treasure Island, so I didn't answer.

"Dudley, get out here or it will go worse for you."' Jake-Jack-John responded, "Choose your weapons!"

It was growing dark, and Mr. Hennie began pulling down his tent.

"Break camp!" Pointing at me, he yelled, "You! I'll deal with you later."

In the process of taking down the tent, Jake-Jack-John escaped. As soon as all our tents were down, Mr. Hennie told us to put them up again.

It grew darker, and we made him help us. From time to time, as he strode anxiously among the tents, he looked like he might burst into tears.

To keep from doing so, he mumbled, "Burn, bash and bury!

Boiled beef and carrots, burn bash and bury." As soon as a tent was put up, he confined its occupants. "To your barracks!"

I shared a tent with another boy who'd been on the hike.

Later that night, we were lying on our backs, staring at the fabric ceiling above us, when we heard someone mutter, "Boiled beef and carrots," in their sleep.

"Sometimes people who were in the war get flashbacks," I whispered. "They call it shell shock."

He turned his head toward me, listening.

"The Nazis did horrible things," I went on, trying to remember exactly what I'd read in a library book I wasn't supposed to check out. "They starved people. Beat them. Made them work until they died. They gassed people. In ovens."

I wasn't sure if oven was the right word, but I said it anyway.

"They did experiments on twins. And if the people were Gypsies, they took their wagons and guitars and made them build bombs."

My voice had gone thin. The tent was quiet again.

Neither of us said anything after that. I think we both just listened to the wind in the trees, not sure what to do with the pictures in our heads.

Mr. Hennie must have heard. When he had composed himself, he stood by our tent and told stories about the war.

He knew Dudley was the "primary offender" and said he would seriously consider reporting this to the group committee. "And that necessarily means your father, Gabriel."

From the trees bordering the camp, Jake-Jack-John yelled, "You do, and I tell everybody your pretty daughter at university poses in dirty magazines. And I'm not a primary offender, you dink. I'm an elementary offender; I'm an intermediate offender. You don't know anything. You don't even know who's taking pictures of your daughter!"

Mr. Hennie was the last to put up his tent. By the time he went to bed, it was dark. If he noticed Jake-Jack-John had peed on his pillow, he didn't say anything.

Jake-Jack-John slept on a rock ledge by the riverbank and refused to go home the next day. He kept to himself and occasionally threw a rock or stick at the tents while everyone else prepared for the morning inspection.

None of the other Scouts wanted to talk about Mr. Hennie's dirty magazines. They had tuned out Jake-Jack-John in the same way they ignored Mr. Hennie striding around commanding us to "burn, bash, and bury."

At two o'clock in the afternoon, some of us started talking to Jake-Jack-John by the big rock on the south side of the river. We were ambushed by older Scouts from across the river. They threw rocks and sticks, and some threw spear ferns.

It was like the Athenian navy invading Sparta, with so many spears that they darkened the sky. A rock struck me in the mouth, knocking me out.

# CHAPTER 9

**Liza Put Out a Beautiful Wash**

To this day, many will still not admit it, but most people realized Jake-Jack-John's mother was the most attractive woman on the Island.

It wasn't just her looks, though they helped. She had that kind of brightness, the way she walked into a room like she already belonged there, smiling wide and laughing loud, never once seeming burdened by life the way other people did.

One day, her great measures of beauty, which included an attractive strength she had about her, a joy in life, an obvious grace, drained from her face. She appeared at our house at dinner time, wanting an immediate meeting with my mother.

Winona Dudley usually wore lipstick and, always, perfume. When she arrived, my mother ran off to change clothes, put on lipstick, and brush her hair.

"Pat! Pat! I need to see you right away!"

Although Winona had never been inside our house, she found my mother in the bathroom and burst in, closing the door and locking it behind her. They stayed there for an hour.

When my mother emerged, she told me to find a serviette. "Mrs. Dudley is going to be here for a while."

Earlier that day, the mill manager and Mr. Hennie had visited

Winona to tell her that Mr. Hennie's daughter was getting married in June. Winona had been selected to choose an outfit for Mrs. Hennie.

Jake-Jack-John had been hiding on the stairs, listening. When his mother asked what kind of budget she could expect for the effort, she was offered a thousand dollars, "if you show receipts."

Jake-Jack-John made his move.

"A thousand dollars! A thousand dollars?"

Mr. Hennie leapt to his feet. "This has nothing to do with you!"

With equal enthusiasm, Jake-Jack-John sprang toward the mill manager.

"A thousand dollars? Are you as crazy as he is? Do you know what a thousand dollars buys for shrink services these days? My shrink at Brannen Lake gets a thousand bucks a week. Mrs. Hennie needs a month of Sundays' worth of shrink work, and that's probably at double time-and-a-half."

Everyone in the room was shocked into silence, not because Jake-Jack-John was there, but because they all knew everything he said was true.

He continued to hold court.

"When did you say the wedding would be? June? Maybe you could swing some kind of deal to pay time-and-a-half because a month's worth of shrink time isn't going to do shit for that poor woman. When was he last time you talked to her?"

"This is enough!!" Mr. Hennie shouted.

The room exploded.

Adults lunged in every direction, some trying to hold back Jake-Jack-John, others scrambling to stop Mr. Hennie from getting his hands on him. The air filled with shouts and a tangle of limbs.

Swearing flew like spit, and someone upended a chair. Not one of the kitchen ones with spindly legs, but a heavy, Swedish Modern lounge chair, made of teak and looking expensive.

Amid the commotion, Jake-Jack-John went down.

Later, the adults would all agree that he had "fallen down the stairs and knocked himself out." The phrasing would be repeated, with careful precision, as if it had been printed on cue cards.

The mill manager drove him to the hospital in his company car.

When Jake-Jack-John came to, the doctor, tight-lipped and tired-looking, quietly told Winona, "He needs somewhere else to be for a while."

And just like that, it was decided, he would go stay with my family until he was well enough to be trouble again.

During that time, Winona and a handful of women from the subdivision banded together to help prepare Mrs. Hennie for the wedding. My mother initially was of the view that neither venture would have any success.

Around the same time, we'd rented out our "little house" to a Coast Salish family, a decision that sent quiet shockwaves down the street. A few neighbors whispered disapproval, not outright, but loud enough to make sure they were heard.

My parents pushed back, reminding anyone who'd listen that Canadian Indians had enlisted for the war in greater numbers than any other group. "They fought for this country when many of you didn't," my father would say, not always kindly.

Eventually, the tone around the neighborhood softened. Liza, the tenant, hung her wash like a page from a catalog; bright, clean, perfectly pinned. And the four children, all wide-eyed and well-mannered, won over the block just by being themselves.

But if my parents were prepared to take a stand to help a family from Valdez Island, they weren't sure any neighbour would welcome Jake-Jack-John living on the hill.

***

## Hospital Food Is Great

When Mom and I arrived at the hospital the next day, the staff were ready for Jake-Jack-John to leave. He'd already stolen from the kitchen and, when we got there, he was racing a wheelchair down the hallway. The matron, a large woman named Mable Clayton, dressed like a regular RN from the neck down. From the neck up, however, she wore a nun's headdress.

"Pat! Thank God! That child needs to leave this ward, this whole hospital, now. Stat!"

Jake-Jack-John said, "No, I really want to stay! Hospital food is great."

He giggled, laughed, then broke into sobs, shaking uncontrollably for five minutes. Only my mother and I tried to comfort him. Matron Mable stood with her arms crossed, her jaw set, loudly tapping her big white lace-up shoe on the polished linoleum.

On the way home, Jake-Jack-John said he had never been a bad kid. Everything my mother had heard about him was a fabrication.

"But now, seeing as how I've been knocked out, I may, from time to time, likely forget who I am, or where I am, and I can't be held responsible if I say or do something someone may object to."

He had heard about medicines at the hospital that could help him, but he preferred not to take them because.

"Who knows? They might give me the wrong kind, and I could end up a real loonie."

My mother asked him if he knew what Buckley's Mixture was.

"That shit!" he replied, sucking in his breath. "I've tried it once and it's the most foul stuff I've ever put in my mouth."

"That's the only medicine we have in the house," Mom said, "and if you know what it tastes like and can remember it doesn't taste sweet, you're not a loonie in any way. But if some morning you can't go to school, if you stay at home, you'll have one spoonful at nine o'clock and another at noon."

Jake-Jack-John stayed at our house for about a month. He went to school all the time, and although I hoped he might forget who he was or where he was, it didn't happen.

His mother's efforts to get Mrs. Hennie ready for the wedding are another story.

**Talking to a Curator**

As with many other problems in the community, complex grief reactions to sudden deaths, illegitimate babies, the mothers formed a study group. This time, the topic was Mrs. Hennie's mental illness, and possibly Jake-Jack-John's.

They gathered books from the library and borrowed pamphlets from the doctor's office, underlining phrases like *"early intervention"* and *"chemical imbalance."* Someone even brought a book about hysteria from the '40s, which most agreed was outdated but still interesting.

The women in the community *began* their 'study group' by inviting a psychiatric social worker for tea. The meeting was set in our living room. Folded napkins, banana bread, and questions about humane treatment and postpartum depression.

Jake-Jack-John and I got home from school earlier than expected. We opened the door to find the air thick with perfume and nervous optimism.

Most of the women agreed humane treatment was the decent thing to do, for most people. But they didn't believe that's all Mrs. Hennie needed.

A few women admitted they'd had "a dose" of postpartum themselves. One confessed to hiding in a closet for three days until her sister-in-law drove her to the hospital.

But none of them had ever thrown furniture or walked into

traffic in her housecoat.

Mrs. Hennie had something bigger, heavier, and far less manageable than what they were reading about. It wasn't just depression. It was a kind of serious sorrow encased in layer upon layer of glacial ice.

What came of the reading and the meetings was that Mrs. Hennie got bathed, they shaved under arms and legs, and got her clean underwear and a bra that fit.

They did her hair every day, though a French roll, the use of various kinds of curlers, and scarf accessories didn't work. Finally, they gave her a henna rinse and arranged for her hair to be cut at Millie's Beauty Salon in Chemainus.

They also started walking her all around town, which took up considerable time. They walked with her, and later marched with her as a strategy to get her fit and tone her muscles. When they let her have a night off or when she was asleep, Winona and the others would come to our house for tea.

My mother fed them while they talked about what they were doing, evaluated what was successful and what was not, and what else they could try.

Swimming was out, and bathing was hard enough, as two women in bathing suits had to be in the shower with her.

My mother came up with the idea of installing badminton courts, and money arrived from the mill to paint the badminton courts on the floor of the community hall. We bought nets and rackets, and Mom taught everyone the rules. However, while they all played, Mrs. Hennie just sat on the side, surrounded by Winona and the other women.

I'd been to a few cousins' weddings, and I knew Mrs. Hennie had come a long way, but she still wasn't ready for a mother-of-the- bride performance. Far from it! Particularly when her exquisitely beautiful and intelligent daughter was going to marry the man she met at UBC, whose father was "a senior vice president in management."

Winona negotiated a bit more money from Captain Hennie and took Mrs. Hennie and some of her other helpers to Victoria to shop. They had a good time despite her losing Mrs. Hennie in the provincial museum and taking a couple of hours to find her.

"She was in the Barkerville exhibit, talking to a curator all the time. Imagine that."

Winona had asked one of the women with them to thank the curator and inquire about his thoughts on Mrs. Hennie.

"What an extraordinary question," the curator responded. He added, "She could work here. You know she told me she used to handle first editions in New York City."

At the debriefing tea, nobody knew what first editions were, and maybe the curator was "a little touched himself." The women concluded that Mrs. Hennie had probably sold the morning paper or the late-night reviews printed after a Broadway show's opening.

# CHAPTER 10

**The Bookmobile.**

They had painted her toes and fingernails and bought her white, open-toed slingbacks. They also purchased a couple of stylish, polished cotton-print dresses and a bright pink cardigan. By this time, women would come to her house to clean, dig around the garden, and prune the wild rose bushes.

I hung around the Hennie house on the day the municipal bookmobile showed up. As people gathered around it, the women working in the Hennie house stopped to chat and told the others to get ready, as Mrs. Hennie would be taking two books on Barkerville to the bookmobile.

Winona and Mrs. Hennie came out the side door, Winona balancing a book on her head and Mrs. Hennie holding her arm. They walked, arm in arm, down the grass-rutted driveway, Winona talking and Mrs. Hennie listening. Just before they got to the group waiting their turn on the bookmobile, Winona stopped and took the book off her head.

"Now go off and be sane to the people on the bus. Don't say 'excuse me' too much. You'll be fine."

Winona turned and walked back to the house.

Mrs. Hennie wore lipstick. She wore her new sweater like a cape with a gold chain holding it together. She had on her new white shoes and walked like a normal person. She even smiled from time to time at the other women watching her. I was at the front of the bus, ready to climb on, and Mrs. Hennie took my hand.

"Jeremy," she said, "shall we see what this mobile library has to offer?"

One of the women responded. "Tut-tut, Mrs. Hennie! Tut- tut!!

It's the bookmobile. It's not the mobile library. It's the bookmobile!"

Luckily, we encountered Dorothy.

"What's going on here? Come up, don't you look lovely! And how lucky to be with Jeremy. I've known three generations of Jeremy's family. You're in good hands. Jeremy, you're here to carry away any book or magazine Mrs. Hennie may select."

Dorothy Cameron was the wife of Colin Cameron, the CCF Member of Parliament for the riding, and she told us that running the bookmobile was the best job a member of Parliament's wife could have.

"Election sign locations! Who reads what? Who doesn't read at all?"

The bookmobile was an old school bus with books along the walls, a magazine section, and a check-out desk. Soft music played, so soft, it had to be carefully listened to in order to hear it.

I got my Milly Molly Mandy book and stepped off the bus.

Winona bowed and curtseyed to people standing around the

Hennies' yard. Jake-Jack-John was coming down the road.

It didn't take long before Jake-Jack-John told his mother she had done everything humanly possible for Mrs. Hennie, and asked rather pleadingly when he could go home. But Winona wasn't finished. In fact, she was just getting started.

"This could be a full-time profession," she declared, already brimming with new plans. She envisioned the mill dragging log booms down to the beach so she could swim laps every morning. "And we'll start a town program, children need swimming lessons." She spoke as if she were forming a municipal committee rather than wrapping up a family intervention.

Jake-Jack-John, sensing his extended stay would not end anytime soon, began acting like a three-year-old. He sulked, he whined.

"I just want to come home, Mom! I just want to come home!"

Winona took us aside and told us that Mrs. Hennie wasn't doing well. Sometimes she acted more simple than anyone could imagine.

I felt they needed to visit, and started off to the Bookmobile to see how Mrs. Hennie was doing.

"Stay awhile," Jake-Jack-John said. "I've got an idea."

"Suppose she's just not ready for the wedding," I said. "She will come hell or high water."

"Just supposing, for a minute, she backslides at the shower and starts fishing in the toilet bowl or something, what are you going to do?"

"Someone will be with her more and more as we work up to the big day."

"The next time you go to Victoria to find what she'll wear at the wedding, do the shops have lots of sizes and falsies and everything?"

"What are you getting at, Jake-Jack-John Dudley?

"Well, I was thinking you might need a double for the pictures. Somebody who could kind of sneak into the line. I'm thinking you can cut out Mrs. Hennie's face and stick a picture of her head on the picture of the double. Someone could be walking her down by the graves in the graveyard. "

"Jake-Jack-John, what are you thinking? Who might this double be?"

"Me!"

Winona and I started to laugh. We punched one another in the arms, then punched Jake-Jack-John, and we laughed so hard, we kept falling over one another trying to stand up. Winona said everyone down at the bookmobile would think we'd caught what Mrs. Hennie had, like it was in the water.

"Maybe that's why she didn't bathe," Jake-Jack-John speculated.

"Do you really think Mr. Hennie is going to be up there at the front of the church with you sitting in the front pew?" I asked

"No. Of course not!" Jake-Jack-John replied. "We'll get a double for him too!"

"Who?" Winona and I asked.

"You," said Jake-Jack-John, pointing at me.

We fell into another laughing fit. This time, someone approached Winona and said she "better get all her charges under control."

Winona agreed.

"Jeremy, go down to the bus and see how Mrs. Hennie is doing."

I ran toward the bookmobile, and when I got on, a woman from the community was trying to take a fashion magazine from Mrs. Hennie.

"What's up?" I asked.

"Mrs. Hennie wants to take this magazine out so she can order something from it," the woman said, obviously irritated, "but I've explained to her, over and over, the magazine is out of date, so it's probably not possible to order the thing she wants."

"What does she want?"

"Well, that really doesn't matter, does it?' "It does to her."' "And obviously to you."

I got the distinct impression this woman deliberately wanted to hurt Mrs. Hennie's feelings, and I didn't know what to do. Mrs. Hennie had grabbed my arm and was digging her nails into it. Her eyes filled with tears.

Jake-Jack-John made it to the back of the bus. "What's up?"

The woman tried to explain again, but Jake-Jack-John cut her off.

"Well, if that's how you act in the bookmobile, if I were you, I wouldn't want to show my snag-faced horse face in here ever again. You just don't know what might happen when you're stepping out of here with an armload of books, that's if you take books out of this library. Or are you just here to gawk?"

Turning to me, he said, "Maybe she can't even read."

Then, to the woman, "Be careful. Some little kid could even run into you with their bike. You wouldn't know what hit you."

Not wanting to risk further confrontation, the woman rolled her eyes and took her books to the checkout desk.

"What's the magazine?"

Mrs. Hennie began to tremble. Then suddenly, like a glass cracking under pressure, she broke into sobs. Jake-Jack-John, startled but not unkind, slipped an arm around her shoulders.

"Where's your hankie?" he asked gently.

Wordlessly, she reached into her sleeve and pulled out a soft, well-used handkerchief.

"Good," he said with mock authority. "Now dry those tears. Tell Jake-Jack-John what's going on."

She let out another round of quiet sobs, her body hitching with each breath. But slowly, the edge began to soften. Her tears slowed. Her breathing steadied.

Jake-Jack-John gave her a gentle squeeze and said, "She's just a snag-faced hag. A nosy, miserable, busybody. Don't waste a single tear on her."

He leaned closer, his voice growing playful. "She won't be at the wedding. I'll make sure of it. Maybe I'll… do something to their car."

Mrs. Hennie gave a wet chuckle through the tissue.

Jake-Jack-John smiled, pleased with himself. "That's better.

Now, what's in the magazine?" he asked, reaching across to pick it up, trying to bring the mood back to something lighter.

One of the pages showed a China doll on a music box. It was entitled "Limited Edition," and the doll was the most beautiful I had ever seen.

"I'd like to get this as a wedding present for my daughter," Mrs. Hennie said.

"That's perfect! Beautiful. The snag-faced hag glommed onto it first or something?" Jake-Jack-John asked.

Mrs. Hennie took a deep breath and shook her head, then looked at us with a vacant stare.

"Or something. Yes. Something. Or something . . ." she said, struggling to think of the next word. She just stared, frozen, at the wall of books, but focused on nothing.

"What do you mean, 'or something"? That's crazy talk. Look at me."

Jake-Jack-John took her chin between his thumb and finger.

"Look at me," he said and smiled at her.

Mrs. Hennie spoke quietly as if she were in another place. "Time present and time past are both perhaps present in time future. And time future, present in time past. If all time is eternally present, all time is irredeemable."[5]

From a distance, Dorothy Cameron assessed that Jake-Jack-John and I were managing Mrs. Hennie. She put on a Puccini aria, "O mio babbino caro," from one of her opera records stacked at the turntable of her portable record player.

Jake-Jack-John and I both knew it in our bones, we had lost Mrs. Hennie. All the effort Winona and the other women had poured into her, the patient work of rebuilding her spirit piece by fragile piece, had been undone by one cruel and careless woman. It didn't take a fistfight or a shouting match.

Just a few sharp words, a look of disdain, and whatever confidence Mrs. Hennie had regained crumbled like old plaster. Now she stood shaking in the back of the Bookmobile, lips moving but saying nothing, her voice reduced to a stream of nonsense. Her eyes were far away. It was as if someone had turned off the part of her that remembered how to be in the world.

If Winona had been there, she would not have hesitated. She would have charged that woman like a storm and made her answer for every word. Winona had built something fragile and beautiful in Mrs. Hennie, and now it lay broken in her absence.

"What do you want to do? Jake-Jack-John asked. Do you want to order it?"

120

Mrs. Hennie nodded.

"Then we'll do it." Then, to me, he said, "Look like you're looking for a book."

I scanned the shelves and, while doing so, positioned myself so I could stare at the people on the bookmobile until they all looked the other way. When Jake-Jack-John was sure he could get away with it, he coughed and tore the page out of the magazine, folded it, and gave it to Mrs. Hennie.

"You're looking so good, Mrs. Hennie," he told her. Lowering his voice, he said to her, "That new medicine looks like it's doing wonders for you. I'm not on any meds myself right now, and you'll get there too. I know it. Darn, you're looking so pretty these days. I guess I'll be seeing you at the wedding. 'Bye now."

When he was gone, Mrs. Hennie began to speak. "The present time, time past remark? The quotation by T. S. Eliot was going to be my thesis. 'Time present and time past are both perhaps present in time future.' I had it all planned. Pages of notes. I was going to study the layers of memory and time, how everything overlaps like reflections on water."

She paused and took a breath. "Then I met Captain Hennie.

That was the end of that. Literature turned into laundry, lectures gave way to luncheons."

Her fingers curled around a magazine, she held it up, as if to make a point. "This magazine is only a month old. Someone could phone instead of writing, you know."

She straightened, put on a voice full of confidence, and began an imagined conversation.

"Hello, I'm calling from Canada. Your order department, please. No, not for a subscription. I want to order the limited-edition doll from page sixty of last month's issue. Send it COD to me at General Delivery, Crofton, British Columbia, Canada."

She smiled at the air in front of her, nodding as if hearing a polite reply.

"What's that? You can charge it to my telephone number? Well, how modern. Yes, go right ahead. My husband, the Captain, would prefer that. He hates it when I go up to the wicket and say, 'General Delivery,' like I'm some lonely wanderer."

She gave a gentle laugh. "Oh, the weather here? Lovely. We live on an island in the Pacific. Vancouver Island."

There was a pause as her eyes drifted to the magazine again. Her voice dropped into a softer tone.

"Still in stock? Wonderful. Thank you. Have a nice day."

Then she added, more to herself than anyone else, "Funny how far a little doll can travel when dreams stay put."

"Do you think you can do it on the phone?"

"I think so. I have the New York phone number."

"How about charging it to your telephone number?"

"Well, I guess I'd have to learn my phone number, wouldn't I? I never phone anyone, and I certainly don't phone myself." She laughed, a deep-throated chuckle. "That would be really crazy."

"How do you know how to talk like that to the magazine on the phone?"

"I used to sell first editions and limited editions in New York."

"You mean there are more dolls like that?" I asked as we left the bookmobile.

She pointed elegantly with her thumb to the woman who had been interested in the magazine. Looking up at the sign that read *Adelaide Street*, she said, "Hm. Adelaide. There are dolls like that all over everywhere. They're quite common."

Then she took my arm.

"Adelaide Street. That's right. We're on Adelaide Street, Mrs. Hennie."

"Adelaide is also the name of a character in *Guys and Dolls*, the Damon Runyan musical. She was one of the Hot Box dancers. Hell, Adelaide loved Nathan Detroit." She laughed. "Good old reliable Nathan . . ."

Oh, really," I said as I glanced at Winona, ringing her hands at the foot of the driveway. "Everything's fine!"

Taking my cue from Jake-Jack-John, I said, "Well, I'll be seeing you at the wedding!"

Later, Winona and Mrs. Hennie found a limited-edition music-box doll, but I don't know whether it was like the one in the magazine. Maybe it was one of her sisters.

Jake-Jack-John and I weren't invited to the wedding. Captain Hennie thought Jake-Jack was up to something and had someone arrange for us to attend a church camp on Thetis Island over the weekend. When no one came to pick us up on the day of the wedding, we knew we'd been taken advantage of.

The wedding went well. Everyone said they could believe that Mrs. Hennie once had been as beautiful as her daughter. Some said she was still a "handsome woman," and hoped she would "keep herself up." She did for a while, then Winona started organizing swimming lessons, and then she left town.

After the build-up to the wedding, it didn't take Mrs. Hennie long to return to her old ways. The last time I saw her, she wore a dress inside-out, had stopped shaving her legs again, and hadn't bathed for a long while. She sat on the sidewalk in front of the post office, eating a lemon pie she'd bought.

It had already been half eaten when she offered me a piece, and flies swarmed all over the aluminum pie plate. She had just collected the mail and had a pile of envelopes beside her, using them as serviettes. They, too, had lemon pie smeared on them.

At least she had learned to wipe her mouth after food covered much of her face. Someone should have given her a plastic fork or a Swiss Army knife, but then, if she had a knife, Jake-Jack-John would probably have gotten her to give it to him for a dead butterfly or a few shells.

# CHAPTER 11

**Portia Rhodes Winthrop**

Portia Winthrop was the resident lawyer, though few in town ever thought of her that way. She arrived three mornings a week in a sickly green Buick that looked like it belonged in a museum. The car was driven by her father, a stooped man with tufts of white hair and the solemn bearing of someone long accustomed to delivering verdicts. Portia never drove.

Neither did her mother, who always sat in the back seat wearing a dark kerchief tied tightly under her chin. She never spoke to anyone. In fact, no one could recall ever hearing her voice, not even in greeting or farewell. It was assumed that at some point she must have said "I do," but that was the only utterance anyone could imagine from her.

Portia's father had once been a judge. His father before him had worn the robe, and his grandfather had served as Chief Justice of the Court of Appeals. The Winthrop name was storied and stately, especially among their extended family in Shaughnessy. But when Portia went off to UBC, her relatives kept their distance. They whispered about her odd ways and politely declined to send her invitations. With parents as strange as hers, they reasoned, what chance did Portia have of being anything but peculiar?

Her eccentricities were small but unmissable. Her mother had never taught her how to apply lipstick or wear clothes that flattered. Teenagers in town, girls still figuring out their own sense of style, could manage makeup better than Portia ever did. Eye shadow and mascara were mysteries to her. She focused instead on lipstick, though she applied it as if with a paintbrush in the dark. It bled well beyond her lips, creeping up toward her nose and down to her chin. The effect was unsettling, as though she had been eating cherries with careless abandon.

Mike McKenna told my father he had asked Portia if she owned shares in the lipstick company.

She didn't give much thought to the rest of her attire, or learn how to wear high heels since she'd had a serious knee injury playing grass hockey for the UBC women's team. Until then, she had been very good at the game. After the accident, she never played grass hockey or any other sport.

Her hair and her hats were the worst of her accessories. She may well have dressed in the dark and sometimes pulled her skirt on backwards. And she couldn't have used a mirror when putting on a hat. There were times, throughout a whole conversation, when she had to brush either her hair or a feather away from her face. The feathers often had lipstick on them, and she sometimes had food in her hair.

When we were about nine or ten years old, I once helped Jake-Jack-John look in her window as she changed into her Girl Guide Leader outfit. He wouldn't let me look, but he couldn't keep from laughing, so I let go of him, and he fell on top of me.

Portia came out with the skirt part of her uniform on and the shirt part flying behind her like a bustle. The top of her lawyer outfit was undone, with her blouse left open while the white lawyer collar stayed fastened. I didn't see anything funny except that she needed to finish getting dressed in her uniform, but Jake-Jack-John started singing.

"Portia, Portia doesn't have a bra-ah-ha-ha. Portia, Portia doesn't have a bra, uh huh. Ha. Ha."

Fuming, she stormed into her office and tried to close the curtain. She pulled too hard, and the heavy cloth, which had never hung properly, dropped and failed to close fully. She came out, yelling.

"If you are ever found by me or anyone else hanging around my office, you will be sorry you've ever seen the light of day. I'll get a No Contact Order pronounced! You will not be able to walk on this street. If you violate it, you'll be put in a provincial youth detention facility."

I thought we should leave, but Jake-Jack-John started singing again.

"Portia, Portia doesn't have a bra-ah-ah-ah."

She began yelling again, but Jake-Jack-John kept singing. And dancing. Then, before long, she was snapping her fingers in rhythm to the song and looked like she had started to enjoy the performance.

She buttoned her lawyer blouse and still wore the shirt part of her Guide Leader uniform like a little train with pockets. We got to talking, and she wanted Jake-Jack-John to go to Tailor's store to buy nail polish for her.

He asked why she would want to get nail polish there.

"Mrs. Tailor has had the same pink nail polish on the shelf for two years. It's cheap! It's not the style anymore. Get something fancy. Modern. Red to match your lipstick. Brighten yourself up, Portia!"

When she offered him money for the nail polish, I felt like I was living in a crazy town.

"Get it yourself!" Jake-Jack-John whooped. "You're big enough and certainly ugly enough."

Then he ran home.

After all that, Portia went to see my mother about finding someone to paint her window so that no one could see in. As she told Mom about Jake-Jack-John, she started to cry, saying it wasn't so much that she was spied on, but that she couldn't stand the boys making fun of her.

Mom said she would see that the boys' behaviour was "rectified," and Portia replied that she would see to the window herself.

Mom agreed to call Winona, and Portia said, in that case, she'd "continue serving with the Guides."

*** 

We had been thinking of returning to New Zealand, so Dad took me to Portia's office to see about getting our passports reissued.

128

I'd never been in such a cluttered place. There were piles of paper everywhere, some held down by rocks and oyster shells. A vase of dead lilacs stood on the counter by the phone. From the stacks of papers and files, it appeared she had a lot of work. A brick on top of a stack of books sat on one of the chairs.

After she busied herself for a while, she looked up and asked my father, out of the blue, "Do you know the peregrine falcon is an 'occasional user' of the forested habitat adjacent to the mill and the ferry wharf?"

"I haven't given it much thought," my father replied, "and I probably wouldn't recognize one."

"There was a pair," she added, "two pairs, in fact, and a Saudi prince wanted a pair for breeding purposes. I made sure they left $25,000 as a surety, and that after they had hatched their young in Arabia, the pair would be returned."

My father told her $25,000 meant nothing to the Saudis, and that she should have asked for more.

This was the best deal she could because the Saudi lawyer felt uncomfortable doing business with a woman.

As she explained the process to get the passports reissued, I grew bored and went out to the car.

When my father returned from her office, he said, "Portia is certainly an extraordinary woman, but she should really have something done with her hair, though it's not my place to comment on that."

"Her hair, at least for starters," I said.

"Do you know she's threatening to sue the company to prevent them from putting log booms in front of Jessie McMorran's beachside auto-court. Jessie had plans for the Twin Gables to become a 'destination resort,' and talk in the town was that when Winona heard about this, she laughed and said, 'For those who resort to assignations!'

Jessie hired Portia because she believed the log booms would jeopardize her business potential. My dad said, "She's already getting more business at the Twin Gables from the overflow of construction crews we can't house in the bunk houses. Nothing against Portia, I wonder whether lawyers meddling in others' affairs do as much harm as they do good in a community like this."

"You let her be the Girl Guide Leader," I replied. "Well, nobody else wanted the job!"

***

I was in the same class as Jake-Jack-John when we studied geometry. I taught him the Pythagorean Theorem by singing the lines from the Danny Kaye song.

"The square of the hypotenuse of a right triangle is equal to the sum of the squares of the two adjacent sides."

I translated that to the simple answer: A squared plus B squared equals C squared.

Jake-Jack-John had just returned from Brannen Lake, where he had been given several tests that indicated his intelligence was within a superior category.

The teachers all assumed, while he was away, that he would have learned something.

Jake-Jack-John disclosed to some of the other kids, including me, that he "hadn't learned shit" because he was always fooling around or getting into fights or being dragged away to talk to his psychiatrist.

He was glad to be home, though he made it clear from the start that he couldn't offer anyone any guarantees. He told them he had even given up the pretense of stealing, as if that had once been part of the plan.

With a weariness that settled deep into his voice, he begged the others to help him in whatever way they could. However, no one quite knew how to help or when to intervene. He was unpredictable, like a spinning compass needle, never pointing in the same direction for long.

Before long, he began acting out in school again. Every time the teacher asked him a question, he responded with something absurd, barely cloaking his defiance in humor. "You don't know this yourself, and you're hanging out your shingle as a teacher?" he said once, loud enough for the whole class to hear.

"What kind of world are we living in, I mean, really?" The class fell silent, unsure whether to laugh or look away. The teacher, caught between frustration and pity, moved on without answering.

Sometimes the teacher would share a long, silent stare session with him and end it by asking, "Do you really want to be in this class?"

To which Jake-Jack-John replied, "Do I really have a choice in the matter, or do you, for that matter? Let's just both try to get along."

From time to time, he raised his hand to answer a question.

When asked about the Pythagorean Theorem, he replied, "Easy as pie! A squared plus B squared equals C squared. Put that in your pipe and smoke it!"

Around this time, all the kids in our row had new pencils. After a week or so, Jake-Jack-John called a meeting, even inviting the girls in our row. He gave everyone a couple of pencils, ballpoint pens, and other various ink pens, some quite fancy. He arranged it so that each student in front of another had the same kind of pen and pencil on any given day.

Those were the days when students sat in rows and had to pass tests back to the student behind to mark them. Jake-Jack-John had devised a system where, if the answer was unknown, the line was left blank. The student directly behind filled it in when the correct answer was read out. Our row started getting better marks than anyone else in the class.

Jake-Jack-John came in one day with a brand-new set of pencils and pens, enough for everyone in our group. He passed them out like some kind of benevolent ruler, beaming as if this gesture would change the course of our academic lives. And in a small way, it did. Before long, even the younger students from the other grades began to copy our method, using Jake-Jack-John's system of color- coded notes and sharpened tools arranged just so on each desk. We would have gotten away with it, too. The teachers, delighted by our sudden focus and improvement, congratulated themselves, convinced they were finally getting through to us. Parents beamed at report cards and classroom visits, never suspecting that their children had rallied around a scheme dreamed up by a boy with a talent for chaos.

But that was the thing about Jake-Jack-John. Just when things were going smoothly in one corner of his world, something else always unraveled. If school was going well, then you could be sure trouble was brewing somewhere else, at home, in the woods behind the subdivision, or in town where he liked to wander and test boundaries. It was as if trouble had a homing instinct for him, sniffing him out no matter how well he disguised himself in success.

He had been upset with Mr. Hennie for not allowing him to attend the wedding. Later, when we were on Thetis, he told people we had "been sent to a penile colony because we were both Queens of the May and in no way welcome at a wedding in a church with decent people."

He often grumbled about his relatives in Ontario, describing a wedding he'd once attended there as if it had been the pinnacle of his social life. The memory was vague, faded like an old photograph left too long in the sun, but it clung to him as something golden, something other kids seemed to have all the time.

"That was the last real party I ever went to," he'd say wistfully. "People laughing, music playing, everyone dressed up, and food just piled on tables." Then he'd fall silent for a moment before adding, "Maybe I wouldn't get into so much trouble if people invited me to stuff like that. You know, weddings. Big things."

I didn't say anything. I had scores of cousins and attended weddings frequently. They blurred together in my memory like scenes from a long-running play; familiar, noisy, filled with the kind of mischief and chaos that made you feel part of something. Jake-Jack-John didn't have that, and I think he felt the lack of it like a splinter in the skin: small, maybe, but always irritating and never forgotten.

How did he deal with this? He managed to get Mrs. Hennie's house key before she was taken to the Crease Clinic at Essondale on the mainland. Whenever there was no one at the Hennie house, he let himself in.

He told me he went through all the proofs of the wedding pictures that Mr. Hennie hadn't sent back to "Fat Ass Charlie," the photographer, because "at the core, Mr. Hennie was a cheapskate." Jake-Jack-John concluded the proofs were actually Charlie's property and that Hennie was as much of a thief as he was.

He arranged the proofs to have "some sense of what went on at the wedding, what the bouquets looked like, what it was like to cut the cake with a knife decked out with ribbons and bows. Then I put them away and left."

One day, while rain heavily poured, and Jake-Jack-John was bored with arranging the proofs, he created a cartoon at the bottom of the fridge door with a felt pen. He drew a picture of a dog, a pointer and underneath, he wrote "Jackie" in bright pink lipstick. Then he snuck out, and no one in the Hennie house noticed or commented on the dog cartoon.

At this time, Rupert Hawthorne was on his way to complete his paper route. Rupert had asthma and owned the best three-speed bikes in town, so he could ride around and get lots of air in his lungs.

Since they'd seen one another, Jake-Jack-John thought the best course of action was to ask if he could go for a spin. Rupert was a bright boy and wasn't going to let Jake-Jack-John use his bike to ride away, leaving him with a stack of papers. He also wanted to know why he had locked the door to the Hennie house.

Jake-Jack-John said he was trying a key that he had found, but Rupert didn't believe him. Jake-Jack-John had been going in looking at the wedding pictures for at least a couple of months and hadn't been caught.

He admitted he had been in the house lots of times, but only to straighten things up and return some of the items he'd taken from Captain Hennie during the time he had been the Scoutmaster. He also offered to help Rupert deliver the papers, but the offer was declined.

That night, Rupert told his parents about the encounter, and his parents told Mr. Hennie, who, in turn, told Jake-Jack-John's parents. It was already known he had a key because "sometime in the last twenty-four hours, a collection of fountain pens has gone missing."

Nobody said much for a couple of days. Most people who heard about the pens thought Jake-Jack John had been up to his old tricks, and his probation officer likely would be called. Someone also called the police, and Jake-Jack-John had to give a statement. He told them about the wedding pictures, how he hadn't been invited, and that he was genuinely sorry he'd hurt Captain Hennie's feelings. He would like another chance because he was getting along so well at home and at school.

When the police asked about the pens, he became noticeably anxious. He said he didn't know anything about pens and that he wouldn't be changing his story. The police didn't believe him, and Jake-Jack-John held out as long as he could, which, as I understand, was about an hour of spinning yarns. Finally, he tried to strike a deal. He would do his best to replace the pens before the end of the next weekend if no questions were asked.

He came to school the next day and told everyone that the test plan would be put on hold for a couple of weeks, as he had to return the pens. Most students gave them back that day, but some were away from school sick, most of whom Jake-Jack-John hardly knew. A few had taken their pencil cases home.

Hearing they were sick, he went to their homes, explaining to their parents that he had come to cheer them up. This made everyone suspicious. Having been caught with the key to the Hennie house was no big surprise, as nothing he did surprised most people. Visiting sick students to "just cheer them up," however, was something new and interesting.

When he didn't appear at school one day, we learned that he had been in juvenile court the day before, then taken to the Brannen Lake

Youth Detention Facility. Students botched spelling and other tests, and younger students in other classes explained to Grade 1 students how the system worked. When it was traced back to Jake- Jack- John, our class received a serious lecture, and people from the school board came to give us test after test. After a while, we had grown exceptionally proficient, even without cheating.

The boys in our row were taken to the office and given the strap, but I wasn't one of them. I refused to put out my hands and demanded that I be permitted to phone my parents. The principal, responsible for the strapping, let me use the phone, and I called my father. He had worked the graveyard shift the previous night and immediately came to the school.

The principal, asked us to sit down. Dad said he'd prefer to stand. Both now standing, they entered into a long discussion about my being in league with Jake-Jack-John and other boys in my row, teaching all the children how to be "worthless cheaters." The principal said everyone else had accepted their "just punishments," and I deserved the same, requesting my father's permission to strap me there and then.

Dad didn't say anything. He knew I had probably helped Jake-Jack-John, but he wasn't going to let anyone commit violence on me.

After a prolonged silence, he said, "You have requested my permission, and I am denying it. Are you going to call all the other boys' parents and let them know you forgot to ask their permission before you strapped their sons?"

This time, it was the principal who remained silent. He was big on staring at children, particularly Jake-Jack-John, trying, usually successfully, to intimidate them. It didn't work with my father.

"I guess this meeting is over," Dad said.

Things were never good between the principal and me after that.

We didn't speak unless we had to, and even then, our words were stiff and sharp around the edges. I doubt my father remembers exactly why I called him to the school that day, maybe the details have blurred in his mind, but I know he hasn't forgotten the way he stared the principal down in that office.

After that, word spread fast. The kids all knew what happened, even if they pretended they didn't. Every student who ever sat in the hallway outside the principal's door wondered what they would have done in my place. Even the girls, especially the girls, said it could have been them. Some admitted they'd never have had the guts to ask to use the phone, much less insist on it. Others started to look at my father with a new kind of respect. He became a quiet legend at school, the kind of dad who didn't yell, didn't showboat, but made it clear with just one look that he didn't scare easily.

That moment did more than just change my standing with the principal. It changed how people saw me, too. I wasn't just the kid who got sent to the office anymore. I was the kid whose dad came down, took one look at the principal, and made everyone wonder what would've happened if he hadn't.

# CHAPTER 12

**Ernie's Birthday Party**

When Jake-Jack-John came back from Brannen Lake, the twinkling lights in his eyes had disappeared. He had a pig-shave haircut, and some parents warned their kids to be careful, saying he had developed a "mean streak."

He had left Brannen Lake around the time Rupert Hawthorne's younger brother, Ernie, was discharged from the hospital after having his elephant-like ears surgically pinned back. Ernie was going to have a birthday party, and all the boys in his class, except Jake-Jack-John, had been invited.

Winona was still around, and she came to ask my mother to call Honoree Hawthorne.

"It would be bad for Jake-Jack-John if he were excluded. And he's very much trying to get along . . . "

She explained that whenever Jake-Jack-John left Brannen Lake, he tried to be good for a while, but as nothing had essentially changed, it was only a matter of time before something occurred.

"This won't be good for anyone, least of all me, if he starts acting out again. Tell her, Pat, tell her this: I'll even buy Ernie the nicest present any kid could want."

I never knew exactly what my mother said to Honoree to change her mind, but whatever it was, it worked. At first, Honoree was more than resistant. Calling her reluctant would be too mild. She wasn't just hesitant, she was opposed, and with good reason.

Honoree was a thoughtful and articulate woman, shaped by discipline and a cultivated sense of right and wrong. She had once been a kindergarten teacher and was also a trained artist, the kind who could look at a child's crayon drawing and discern the nuance that others missed. Her home was an extension of that same aesthetic, neatly arranged, softly lit, full of things that held both beauty and memory. She had every reason to be wary of Jake-Jack-John.

Her concerns weren't abstract or hypothetical. The last time he had been in real trouble, it was Rupert Hawthorne who saw him slipping out of the Hennie house, and it had landed Jake-Jack-John back in Brannen Lake.

That was not so long ago. Honoree had to consider the risk: to her family, to her possessions, and even to her cats, which she loved as if they were people. Could she trust him in her home, around her child, during what was meant to be a peaceful, joyful gathering?

But Honoree was also a woman of faith. She was active in the altar guild at the Anglican Church and took seriously what it meant to live as a Christian. Forgiveness wasn't just a word to her; it was a discipline, a kind of spiritual practice.

In the end, perhaps that was what tipped the balance. Not the pleading or persuasion of others, but her own belief that even someone like Jake-Jack-John, for all his failings and flare-ups, might deserve another chance.

And so she said yes. Quietly. Cautiously. But yes, all the same.

"He can come to the party," she told my mother, "but at a cost. I want to hang my art in the community hall, and you must help me do it."

Honoree got to hang "one piece of her art at any time, anywhere in the hall," and Jake-Jack-John attended Ernie's birthday party. It was a weird event. We were supposed to play hide and seek in the bush behind the Hawthorne house, then have a turkey dinner. I hadn't seen Jake-Jack-John much since his return, and he was now in a class behind me. After we played at playing hide and seek, we asked Honoree if we could help in the house since it was so cold outside.

Jake-Jack-John said, "I'm worried I could catch a terrible cold." He stopped and shivered for five seconds, then lowered his head and murmured, "I won't be any trouble, Honoree."

Honoree didn't want us "helping out," but she gave us pencil crayons and fresh newsprint from the mill and said we could go downstairs to the basement to color. The cement floor was colder than anything outside, and Jack-John started drawing a picture of Ernie with his paper birthday hat held up by two big ears.

In a second picture, he drew Ernie with his eyes poking through the equilateral triangular points of his tissue birthday crown. The tops of his flattened big ears poked out of the crown that rested on his shoulders. Under the first picture, Jake-Jack-John wrote, "Ernie, eleven years old."

Our laughter broke the silence. Good old Jake-Jack-John was still alive.

Honoree must have heard us, and she asked, "Any monkey business going on down there?"

"No, we're fine," Jack-Jack-John replied. "But I'd like a few more pencil crayons or any other art supplies you can spare. I've been taking classes in art enrichment and expression. If I ever grow up, Honoree, I'd like to be an artist like you. Maybe I could even be a printmaker or a commercial artist. My psychiatrist thinks I have expressive as well as technical skills."

"Interesting," Honoree said. "We should talk about your art interests sometime. I can show you my pastels and my studio. But I don't have much time," she yelled down to us from the kitchen.

"The medium I'm working in today is food as an art form, specially cut vegetables, carrots and yams together, cut with a wavy cutter, Brussels sprouts mixed with savoury stuffing, and red peppers. The mashed potatoes are shaped into the letter E for Ernie, then grilled on a baking sheet, so I'm busy."

I didn't go into the studio, and I didn't want to go outside. I felt rejected by Jake-Jack-John and wanted only to sit and mope. And I said as much to Honoree.

"Cooking a turkey dinner for fifteen boys and being creative requires focus," she said. "While I'm icing a cake, you can sit on the kitchen stool and watch and talk, but don't touch any of my cooking implements."

While we talked, Jake-Jack-John got a handful of pencils and other art supplies and then went downstairs again, waving a large felt pen. I thought nothing of it at the time.

Ernie's birthday party marked more than just another year gone by. It was, in a way, his own soft debut, a quiet reintroduction to us all. He had received a generous collection of gifts, the kind that made it clear things were changing in his life.

This was a party that gave him, and maybe all of us, a sense of affirmation. A fresh haircut, a new shirt with pressed collars, and that unmistakable grin told us he knew it too. Ernie wasn't the same boy we used to tease for the missing button on his jacket or the scuff on his shoes. Something had shifted.

And for the rest of us, we were standing on the edge of that shift ourselves. We were nearly twelve, brushing up against adolescence, but still untouched by its full weight. Not yet teenagers with all the drama and brooding that came with it, we were perched at the summit of elementary school, a position we didn't take lightly.

We were the oldest boys in the building now, the ones who got nods in the hallway and called first at recess. And while we still clung to some of our silliness, a new awareness was also forming. We were watching everything more closely, feeling the change in the air and wondering what kind of men we'd turn out to be.

Honoree surprised us all that day. We had expected sandwiches and store-bought cupcakes, maybe a pitcher of Kool-Aid on the table. Instead, she presented food as if it mattered, with sliced vegetables arranged like a painter's palette, baked potatoes sprinkled with parmesan, and mini skewers featuring glazed sausage and pineapple.

She even talked about food the way a curator might speak about sculpture. And to our surprise, we listened. Something in the way she spoke made us feel older, like we'd been allowed into a room usually reserved for adults.

Change was in the air outside the party, too. Thanks to successful union negotiations at the mill, many of our families were riding a new wave of prosperity. There were freshly built houses along the edge of town, bungalows with wide picture windows and pristine front lawns.

More and more driveways now held a Pontiac Parisienne station wagon with push-button windows or a sleek black Chevrolet Impala. For boys like us, these cars weren't just transportation; they were symbols of status, proof that our families were stepping into a new era of success.

Of course, our mischief didn't disappear entirely. Honoree's baked Parmesan potatoes in the shape of an "E" turned out perfectly; she had used a pastry bag to create the letter on a cookie sheet then baked them. It was only a matter of time before someone bit off the bottom of the E to form an F. Someone else ate the middle integer and turned on its side, voila, a U, and turned upright, a C. Then with chewed off pieces, a K.

"This is a first!" Jake-Jack-John said. "When I get home, and my folks ask how the party was, I'll say Honoree made us eat fancy French things called spud-fucks."

Honoree started to yell at all of us, but mostly at Jake-Jack-John. "What kind of dirty little deviant are you?"

We responded by gobbling up the evidence, at which point it was time for the cake.

As Ernie was about to blow out the candles, Rupert had a bad asthma attack and had to be rushed to the hospital. We all helped cut the cake and wrap it in wax paper so we could take it home and eat it.

Three or four of us walked home together, and some of the boys waited for their parents to pick them up. Jake-Jack-John was again silent but, from time to time, laughed. Otherwise, he was uncharacteristically silent.

A few of us asked, "Come on, tell us what's so funny."

I volunteered that it involved the before-and-after picture of Ernie. Jake-Jack-John just laughed.

Besides her art and being a member of the altar guild, Honoree was proud of her large Indian rug in the living room, her stylish Persian lamb coat, and her new deep freeze downstairs. When Jake-Jack-John went to get more art supplies, he got his hands on one or more felt pens and pink nail polish.

When Honoree, or anyone else, went downstairs to freeze the turkey leftovers, they didn't notice the pink dog cartoon on the bottom of the freezer. It went unnoticed until midday Saturday, when Honoree saw it.

Jake-Jack-John had outdone himself. This time, besides a felt pen, he dressed the dog in a pink suit with a piece of Honoree's

Persian-lamb coat as a collar. Underneath, he wrote, "Jackie."

Honoree called the police. She also called my mother, screaming that Jake-Jack-John should be put in a reformatory that afternoon and stay there until he was an adult, maybe even beyond then. My mother repeatedly said how sorry she was that whatever had happened had again involved Jake-Jack-John.

But she stressed that it was probably best to let the authorities manage the situation. Honoree continued screaming and swearing. At the time, Mrs. Hennie was visiting, as she was out of Essondale on a "temporary visit," which was part of her discharge plan.

"Just hang up," she told my mother. "Just hang up."

When Mom held the phone away from her ear, Mrs. Hennie walked over to the wall and pressed the hang-up button.

It would have been wise to simply leave the whole thing as it was and let the authorities handle it. My mother, however, insisted that we go to see Winona and make sure she was warned that if anyone might do Jake-Jack-John harm, Honoree was close to it.

We tried calling the Dudley home, but their line was busy, likely with Honoree yelling at them, so we piled into the station wagon. Jake-Jack-John was nowhere to be found.

Winona and some of her neighbours were preparing for a cocktail party. They had all heard that Honoree was upset, but there was nothing anyone could do about it. Someone had seen the damage to the fur coat, and Jake-Jack-John had taken a piece of the fur already ripped away from the lining.

"About the deep freeze," Winona said, "well, yes, felt pen doesn't come off, but she's an artist. She could have tried turpentine or linseed oil or apple cider vinegar, whatever artists use."

Honoree and Mr. Hawthorne appeared just as we were about to leave. To say they were upset would be an understatement. Winona graciously told them she was sorry, "and if there is anything I can do, if you could ever use a hand with anything, don't hesitate to call or just drop by. The door is always open."

She also wanted them to come in for "cocktail party goodies, and we can have a nice drink and talk about this like adults."

It was hard to hear what Honoree said, but the gist of it was that she wanted nothing to do with Winona, her son, her family, or her friends. She just wanted her to replace the deep freeze and her Persian-lamb coat.

People had been complaining to Winona about Jake-Jack- John for at least ten years, so she knew how to handle the Hawthornes.

"I think you need to talk to his probation officer. Here's his card, and here's the psychiatrist's number too."

Honoree took the cards and tore them up in front of everyone, gawking at her from behind the sliding glass door. She reached into her purse and pulled out the before-and-after pictures of Ernie that Jake-Jack-John had drawn.

"I'm taking these to his probation officer, or his psychiatrist, or the judge, or whoever else you suggest. We're getting a lawyer, and you haven't heard the end of this!"

As Honoree and Mr. Hawthorne moved to leave, Winona began shouting.

"Just hold it! Hold it there one damn minute. What's that thing you're waving around? Some new expression of your talents? You want me to see it, or don't you?"

As mad as she was, Honoree panted for breath. She stamped her feet, and her face turned red. I thought she might be on the verge of a seizure or a stroke.

Between gasps, she spat out, "This is not my artwork! This was done by your twisted, *twisted* child!"

She threw the drawings at Winona and collapsed against Mr. Hawthorne, who struggled to hold her up.

I had never seen an adult throw a tantrum before. Jake-Jack-John could upset Mr. Hennie, but his anxiety was nothing compared to Honoree's outburst.

I had seen adults cry loudly at funerals, but Honoree was doing a fair bit more than having a good cry. She sobbed uncontrollably, and Mr. Hawthorne could barely keep her standing.

Winona and my mother, along with a few other women, tried to comfort her, but she slapped at their hands and even tried to kick them.

"To hell with this!" she said. "Get a lawyer. Come on in, everyone! I've had enough of this song and dance."

"Give me back those pictures, you hussy!"

At this point, to break the tension, some of us watching began to laugh. No one called Winona a hussy. We exchanged confused glances. What *is* a hussy anyway?

"What did you call me?"

"You heard me loud and clear. Hussy! Floozy! Or is there something wrong with your hearing? Is that another burden the good Lord has visited on your wretched family? Give me back those

sketches this minute, hussy. They're part of the lawsuit!"

Winona signalled to the rest of us to come and view Jake- Jack-John's artwork.

"Part of the lawsuit? Sketches? They're sketches! Okay, let's see if my boy has talent."

I thought I knew what was in the pictures. I really did. But I hadn't realized Jake-Jack-John had taken it upon himself to add more detail, flourishes that turned them from mere mischief into a spectacle. Winona was the first to see them and burst out laughing, a sharp, irrepressible sound that quickly pulled others into her orbit.

Soon, the whole group was doubled over, giggling like schoolchildren.

Honoree, however, was not laughing.

She lunged for the stack of prints, her face a crimson mask of fury and embarrassment. Winona saw her coming and darted inside, slamming the sliding glass door shut behind her and twisting the lock with theatrical flair. From the deck, Honoree banged on the glass with the flat of her hand, shouting for Winona to open the door, her voice rising in pitch. But the sound of laughter inside only grew louder.

It was unfortunate timing, guests were just starting to arrive for the cocktail party. They stepped onto the deck, expecting canapés and polite conversation, only to be greeted by the sight of Honoree pounding on the door, her dress wrinkled, her face wild with indignation. There was no graceful way to explain it, and even if there had been, she couldn't get a word in before the laughter overwhelmed any sense of order.

Eventually, someone must have called the police.

They arrived unsure of what they were walking into. Honoree, breathless and trembling, couldn't form a coherent sentence. The officers glanced around at the still-laughing guests, who tried to suppress their amusement but failed spectacularly. The situation teetered between comedy and concern until one of the officers made the call for an ambulance.

Honoree was taken to the hospital. No one said it out loud, but everyone understood, it had been too much.

Later that evening, Rupert Hawthorne suffered a serious asthma attack and was rushed to the same hospital. He and his mother remained there through the weekend. By Monday, the story had already started to pass into legend, reshaped and retold at coffee shops and card tables, with each version a little more outrageous than the last.

Honoree had sounded rational when she arrived at the hospital.

She explained to the doctor what had gone on and said that for the life of her, she couldn't understand what Jake-Jack-John had done, or what his mother said that had caused her such distress.

The doctor empathized with her and discussed how the behaviours of a child like Jake-Jack-John could cause even the most capable adults to experience frustration. Honoree discussed how other children in her career as an early childhood educator had caused her sufficient concern to have referred them for various forms of counselling. They all had improved.

The situation at the hospital took a darker turn later that evening. While the cocktail party buzzed with music and laughter, the Churchill family, next door neighbours to the Dudleys were dealing with a crisis of their own. Their young daughter, a quiet child prone to curiosity and solitude, had somehow wandered upstairs and gotten into Jake-Jack-John's medicine. No one noticed right away. The adults were deep in conversation and enjoying drinks, while the older children played outside under the string of lights.

By the time anyone realized something was wrong, the little girl had collapsed. She was unresponsive and pale. Mary Churchill screamed for someone to call an ambulance, and in the ensuing chaos, Winona took charge. She rode with Mary and the child in the back of the ambulance, keeping her arms around both of them. The flashing lights of the emergency vehicle cast long, jittery shadows on the lawn as they sped away from the party and toward the hospital.

Word spread quickly, and it did not take long for the tension to reach a boiling point.

At the hospital, as nurses rushed to stabilize the child, Honoree caught sight of Winona in the waiting area. Her fury reignited. She crossed the floor in a few fast strides and launched into another attack, accusing Winona in front of staff and strangers alike. Her voice cracked with exhaustion and rage, and her hands shook as she pointed at her. The confrontation escalated fast enough that hospital security had to step in and separate the two women.

A nurse administered a sedative to calm Honoree down. She was taken to a private room, where she remained overnight, lying in a chair beside Rupert. He had grown quiet again, curled under a hospital blanket and wheezing faintly as he slept. Honoree sat beside him, her fingers gripping the armrest, her eyes fixed on the wall.

Whether she was watching it or simply staring through it, no one could tell.

She did not speak again until morning.

# CHAPTER 13

**Laughing Our Heads Off**

On Monday morning, Jake-Jack-John arrived at school with his sketches tucked under one arm like precious scrolls.

With dramatic flair, he selected a wide patch of the stucco wall, pockmarked and gray from years of soccer balls and rain, and began taping up his drawings in a straight line. Each piece was placed with care, forming a kind of makeshift exhibition.

By the time the last sheet was up, a small crowd had gathered, curious and already giggling. Jake-Jack-John did not rush. He pulled from his coat pocket a long, pointer-like stick and tapped it once against his boot, like a conductor calling the orchestra to attention.

Then he began.

He gestured to the first sketch, the crowd leaned in, whispering and pointing. At first, it was hard to tell what he had added. The drawings were based on Ernie's before-and-after pictures, the ones he drew during the birthday party when Honoree was upstairs and Jake-Jack-John had free range of the basement.

Now, bodies appeared in the *sketches*.

"You all know Rupert and Ernie and his mother are always going on about how they're related to royalty. Well, I'm not disputing that, but if you look closely here, you'll learn something about royalty."

We all stood there, clustered around the stucco wall, watching as Jake-Jack-John pressed a finger against one of his sketches and began to speak with a grin spreading across his face.

"See these little pencil marks?" he asked, drawing our attention to a faint detail we hadn't noticed before. "That's Ernie's penis. This first sketch is the 'before', before his operation. And here's the 'after.' In the 'before,' if you look really close, you'll see he's got two little testicles still up in his stomach. Right here," he tapped twice. "Now, in the 'after' picture, here they are, down where they're supposed to be."

A ripple of laughter passed through the group. It started with a few snickers, then escalated into loud bursts of hilarity. But Jake-Jack-John raised a hand with authority.

"Shut up. There's a whole lot more to know than what you've seen so far."

We all quieted, eyes fixed on him. His tone had shifted, part medical lecturer, part street-corner prophet.

"Ernie's going to be fine," he continued. "He was lucky. He had a good surgeon. They did his ears and his nuts all at once. Pretty efficient, really. But sometimes, it doesn't go so smooth.

Sometimes, if you wait too long, they go in, even with all the modern equipment, they can't find the testicles. They just disappear."

He paused, letting the silence stretch just enough before delivering the punchline.

"That's what happened to the Queen's uncle. That's why he couldn't be King. People say he ran off with an American woman who collaborated with the Nazis, and get this, she had two sets of plumbing if you get what I mean, but the real reason was he couldn't father kids. Because he didn't have any nuts."

There was a gasp, followed by a burst of laughter, a few voices shouting over each other.

"You're full of shit!"

"Where'd you hear that?"

"You're lying."

Jake-Jack-John just grinned. "I read it in a STRICTLY CONFIDENTIAL medical report written by Sir Philip Livingstone, Surgeon and General Family Practitioner to the King. I got it from a guy who knows about these things."

Some of us were still laughing. Others were staring at the sketches with new eyes. Then Jake-Jack-John leaned forward, lowering his voice just enough to draw us in again.

"And that's not all. You've heard by now that Ernie and Rupert's mother had to be taken away on Saturday? Yeah. She'll be joining my friend Mrs. Hennie over at Essondale. But between you and me, Mrs. Hennie will probably get out first. That lady's tougher than she looks. And Rupert? He had to go to the hospital too.

Anyone know why?"

He stepped back from the wall, folding his arms and surveying us as if he were both teacher and ringmaster, letting the weight of his words settle into the cool morning air.

Someone interrupted. "Asthma attack." A few kids nodded as a teacher approached.

"Asthma attack?" he exclaimed. "Asthma attack? That's what you think? Anyone here care to guess why they took Rupert away and why we may not see him for a while? Not asthma. Not his ears. Rupert has quite nice ears if I say so myself. No, Rupert doesn't have any nuts. Class is over. Somebody help me with these sketches. Quick!"

We weren't fast enough. A teacher, then the principal, obtained the sketches. Jake-Jack-John was sent home and, much to the disappointment of the Hawthorne family, he wasn't sent back to Brannen Lake.

It was said he had rifled through a confidential medical records cabinet there, and the staff couldn't guarantee the security of information about any other child who had been admitted.

Jake-Jack-John went to a foster home in another town and, I presume, went to school there. Before being taken away, he asked his neighbour, Mary Churchill, whose daughter had swallowed his medication, to contact Portia, as he would need a lawyer for child welfare court.

When Rupert came back to school after lunch hour, it wasn't long before everyone wanted to have a look at his scars.

"What scars?" he asked.

When he learned the story, he was so mad at Jake-Jack-John that he went and punched a kid he thought lived in the subdivision. He beat up little Donald Schmidt, whose parents were socialists and United Church members.

From time to time, kids teased Rupert about not having testicles, and it was rumoured that older boys made him drop his pants to prove he had both of them.

Mr. Hennie divorced Mrs. Hennie and moved to Ottawa, where he became a policy analyst in the Department of Veterans Affairs. Before he moved, Mr. and Mrs. Hawthorne bought the Hennie deep freeze and moved the original one into the yard, turning it upside down so that children couldn't crawl in and suffocate.

One winter, there was a heavy snowfall, and the Hawthornes had their driveway plowed. When the clearing was finished, the Cat operator accidentally pushed the deep freeze into the deep ditch in front of their house. Then it snowed for another two weeks.

No one noticed the deep freeze was gone until the snow thawed, and the ditch overflowed, flooding a neighbour's garage, workshop, and basement. When the temperature dropped again, the deep freeze blocked the flow of water in the ditch, causing ice patches to form on the road. A couple of cars skidded and crashed.

A forklift was brought in from the mill to move the deep freeze to the dump. In the spring, I went to find it, but the dump managers had attempted to burn and then crush it, and I never saw Jake-Jack-John's art again.

After they bought Hennie's fridge, Honoree told people there was another "pink lipstick cartoon on the door." No one believed her, thinking the incident with Jake-Jack-John had affected her.

Mrs. Taylor from the store, who knew Honoree from the altar guild, suggested she use nail-polish remover and gave her several bottles free of charge.

"Just consider it a lost litre, dear. I'm sure it will do the trick."

If Honoree had talked with her wise, not easily ruffled, altar guild friends before she confronted Winona, a lot of unpleasant things may not have happened. She could have kept the sketches, Rupert wouldn't have been repeatedly made fun of and humiliated at school, and poor Donald Schmidt may not have been beaten up.

She could have used nail-polish remover, or she could have left it alone, or cut off the bottom corner of the deep freeze with a blow torch and called it a modernist statue.

Most people who remember any of this consider it a contradiction. Honoree was intelligent, insightful, and cultured, and had a good heart. Why did Jake-Jack-John's vandalism trigger her so much, and why did she let him play in the basement by himself? Think of what he could have gotten up to down there.

Portia didn't represent Jake-Jack-John in family court. While renegotiating the surety for the loan of a pair of peregrine falcons to the Saudi royal family, she contacted a senior lawyer in the provincial Ministry of Energy, Mines and Resources.

He was an older gentleman, and when they spoke on the phone, he mentioned being aware of her family. He considered it an honour and a privilege to assist her in renegotiating the surety with the Saudis.

Although he stressed that he was never one to mix official government business with pleasure, he asked Portia if she would be amenable to meeting him for dinner in the Bengal Room at the Empress Hotel.

She had the good sense to track down Winona and ask if she could assist her in dressing for a date. Winona hurriedly cancelled her appointments and spent five days with Portia, taking her to have her hair and nails done, then taking her shopping.

The result was a stylish bouffant hairstyle and a smart gold brocade dress with a scalloped collar, sheer sleeves featuring satin cuffs, and a bolero. Portia wanted to wear pearls, but as it was a first date, Winona lent her flashy costume jewelry, which Portia thought looked cheap.

Nevertheless, she was proud of her outfit. On the night before the date, she came to our house to model it for my mother. They ended up making a necklace out of round Lifesaver wrappers, varnished, and held together with sparkly silver-elastic string.

Between each wrapper, there were two or three Lifesavers, also varnished, so that Portia wouldn't chew on them if she got nervous, and, miraculously, she looked like a well-groomed, wealthy lady lawyer.

She leaned in toward her date with a teasing smile and said, "I'm always on the lookout for craft projects for my Girl Guide troupe." As she spoke, she began unfastening the candy necklace she had worn around her neck all evening. She slipped them off and then, almost absentmindedly, wrapped them twice around her wrist like a bracelet.

From the neckline of her dress, she reached in with a practiced elegance and drew out a strand of pearls. They gleamed softly in the low light of the room. "These," she said, holding them up with a quiet pride, "are real. Not cultured."

She let the statement hang in the air a moment before adding, "They were a gift to my great-grandmother from Prime Minister John A. Macdonald himself. She helped him win the nomination when he ran in the riding of Victoria. Apparently, he was so grateful, he sent these from Ottawa with a handwritten note."

She clasped the pearls around her neck again, her fingers deft and sure, then looked at him, her expression unreadable except for the glint in her eye that suggested she had more stories where that came from.

The date went well. Somehow, between the pampadams and the crème brûlée, the aging lawyer and Portia became smitten with one another. Within a matter of weeks, she married the most senior lawyer in the Department of Energy, Mines and Resources.

Shortly after the wedding, which had been a modest family affair, she received word from a relative that an elderly cousin in Shaughnessy was slipping into dementia and had reportedly begun giving away family treasures to hippies. At first, she was skeptical. She insisted that the only things of true value in the family were her own pearls and perhaps a few pieces of antique furniture.

She doubted that hippies would have any interest in such items.

After all, she had read Joan Baez's *Daybreak* and was convinced that those who followed the countercultural path lived by Gandhi's ideals. They believed in non-possession and non-violence, she explained. "Moksha and ahimsa," she added, as if she were reciting the tenets of a philosophy she had carefully adopted, or at least admired from a distance.

She went to Vancouver to care for her cousin, and, within a matter of days, the cousin died by suicide after hanging herself in the basement with a piece of hemp rope. Portia was charged with her murder.

Her lawyer informed her that the evidence was circumstantial and that the charges would be dropped. Instead, she was convicted.

Shortly after her incarceration, her husband had "a neurological incident," and he died during surgery at Royal Jubilee Hospital. Within a matter of months, Portia went from single, small- town lawyer to having a lawyer husband with a senior position in government, to widowhood, to being a convicted murderess.

Throughout her incarceration, and until her death a few years ago, she maintained that the charges against her, her conviction, and even her husband's death were connected to information they had uncovered, an obscure BC Hydro document entitled Vancouver Island Thermal Coal Generating Assessment. Portia believed BC Hydro was working with Atomic Energy of Canada Limited to place a nuclear reactor on the forested land between the mill and the ferry dock.

Her conspiracy theory proposed that the assessment for a coal-fired facility included federal and provincial entities strategically planning to situate a deuterium CANDU heavy-water nuclear reactor on the land where, years earlier, Portia and a member of the Saudi royal family had captured two peregrine falcons.

Portia had no friends in the provincial government. On one occasion, she stated that "Child-welfare authorities should investigate a BC Hydro muckamuck, Robert Goodhour, and his wife for naming their daughter Boobs." This was a sensitive matter, as Portia was quite well endowed, " and the idea of objectifying a teen girl on one physical characteristic would likely cause her irreparable psychological damage or worse."

One year, when Portia was in the Twin Maples Correction Facility near Ruskin on the Lower Mainland, my mother, then secretary-treasurer of the Vancouver Island Quad-Zone of the Royal Canadian Legion, went to the National Dominion Convention. As the minister of justice at that time was a veteran, my mother wrote to him.

Please give this matter your early attention. The above-mentioned is known to me, and I have had her in my home on several occasions in the past.

She is at present, as you are no doubt aware, serving a life sentence and is at this writing in the Twin Maples Farm in BC. I am going to the Legion Dominion Convention in Penticton this weekend and hope to visit Portia en route.

It appears to me that her case should be brought to a new trial, and I hope you will give this matter all consideration within your powers as Minister of Justice.

Yours fraternally

I still have the carbon copy of my mother's original note. Because of the appeal to her Legion comrade, who happened to be the federal minister of justice, and Portia's election to the federal Women Prisoners' Union, I was permitted to pick her up at Twin Maples.

We rode to the Vancouver airport, where she boarded a flight to Ottawa to attend meetings and interview women prisoners at the notorious federal Women's Pen, P4W, in Kingston, Ontario.

Portia looked like a completely different woman. She wore a chic suede suit with a front kick-pleat and carried a leather briefcase that matched her sleek, knee-length boots. Her hair had been French-braided by one of her fellow inmates, and she wore a soft, light-colored lipstick that suited her surprisingly well.

As we moved through the airport, I walked beside her, mindful of the agreement I had signed. I was to accompany her all the way to the gate, stay there until she boarded, and remain until the plane took off.

While standing in the terminal, I happened to spot a couple of my old social work professors walking by. One of them gave me a curious look and, just as I turned to leave, called out to ask, "Who's the politician you were with?"

"Portia, a union executive and old family friend. She's on her way to Ottawa to meet with a minister colleague of my mother."

I lowered my voice. "Old money. Her great-grandfather nominated John A. when he ran for parliament from Victoria."

And Jake-Jack-John? There were other chapters in his life, some of them difficult, others impossible to forget. But those stories remain tucked away for now, waiting for a time when it might feel right to speak of his antics as a young man and the choices that shaped him.

# CHAPTER 14

**Dinner with a Dietitian**

Billy and JH attended the same church services as we did and invited my father, brothers, and me to dinner after the birth of my sister. There had been a baby shower in town for my mother and sister that afternoon at the church hall.

That night, my mother's women friends, Vivienne, Lydia, and Ruby, came over to watch World War II movies and listen to Lydia's stories about how the Gestapo came and took her "right from a church."

They had suspected she carried messages for the Dutch Resistance, which she occasionally did, but she had nothing incriminating on her when she was caught and stripped. They kept her for three days, took her clothes so that she couldn't escape, and gave her one blanket. When she was released, she knew they watched her wherever she went, until the Canadians came and drove the Nazis out of Holland.

My father had been part of the Canadian forces that liberated Holland. He and a small group of Canadians kept a larger squad of Nazi SS members under guard in Groningen, preventing them from escaping to the islands off Friesland.

Vivienne's husband came home from World War I shell-shocked and suffering from having been gassed. She worked with a CCF politician, Stanley Knowles, who fell in love with her, but they were unable to marry. Her husband was in a veterans' hospital and didn't know who she was, but that wasn't grounds for obtaining a divorce.

Stanley gave Vivienne pearl-and-diamond earrings and a strand of pearls, and she left Manitoba to come to Crofton. When she learned her husband had died, she met and married Harold, the local Fuller Brush salesman.

Vivienne was the secretary of the local CCF Association. On the night of the shower, she told the women about making a Will so that the pearls and diamonds would go to my sister. I loved watching World War II movies, but I'd heard Lydia's and Vivienne's stories several times and felt justified going to dinner.

It was Billy and JH's job to entertain us for the evening. When we got to their marvelous house on the cliff overlooking Sthi'xum, Dad parked the new Pontiac Laurentian station wagon as JH waited in one of the gardens to meet us. Billy, wearing a stylish, bright yellow sundress, waved from the kitchen porch like an excited sunbeam.

She disappeared into the house, then reappeared, coming down the stairs carrying a large tray of glasses and a pitcher of lemonade. We were invited to sit on cushioned wicker lawn chairs in the "shaded garden," where we heard about my sister and how she was "born at the top of the ninth inning with the bases loaded."

Billy interjected. "And with the clock ticking, a town metaphor for having three children of one gender and a fourth of another."

We didn't use words like "metaphor" or "gender," but I sensed it was related to a metronome and thought Billy was referencing a clock ticking, possibly in relation to my mother's age. Mom was forty, and most women in town were done giving birth by then. Some were already grandmothers a few times over.

Cynthia was born at a high point of excitement during a local Little League game. When my father drove up to the ballpark, parked the car, and ran to the bleachers, everyone except the batter, the runners, and the pitcher paid no attention to what was happening in the game. There were shouts of, "What is it, Fred?"

"It's a girl!" Dad said it only once, and everyone cheered, keeping up the cheer as they ran to offer their congratulations.

No one noticed Mouse Vanderheide step up to the plate. The crowd was too focused on my dad's announcement. Then, crack! A clean home run. Hans Doliwa was already on third. Lothar, his brother, nicknamed "Fleet of Foot," was poised on second, and Wallace John held firm at first. With one swing, Mouse brought them all home.

Later, Billy and JH caught word of the play, a highlight from a home game victory against Chemainus. That night, the local pub exploded with celebration, drinks flowing as the winning run replayed on every lip.

The lemonade was gone by the time we finished retelling the birth-announcement stories. We were shown around the front and back yards, the vegetable and herb gardens, and the rose garden. As we entered the front door off the verandah, I stared at the planters of huge salmon-coloured geraniums, with nasturtium leaves poking outward, and at the heavily scented bright blue flowers. "What are the blue flowers?" I asked Billy.

"Heliotrope. They're the cilantro of all scented flowers."

I nodded, as if I knew what cilantro was. To make conversation and to discuss something I was interested in, I complimented Billy on her dress.

"That is the most stunningly beautiful sundress I've ever seen. Better even than the dress Princess Margaret wore at the ballpark reception in Duncan!"

Billy smiled. "I'll pass that on to my cousin, Phyllis. She knew Her Royal Highness as a little girl when she visited her grandmother, Queen Mary, at Badminton, where she stayed at the home of the Duke and Duchess of Beauford throughout World War II. Phyllis's father was the groundskeeper at Badminton, and the Beaufords sent Phyllis to London to be trained as a secretary. When Queen Mary was sent to them during the war, she chose Phyllis to be the Queen's companion for the duration, likely to the chagrin of the Duchess, who had hoped Phyllis would play the role of her assistant."

I nodded. What would I know of what a Queen's companion or a personal assistant would do, for that matter?

"And thank you very much for the compliment. I had the dress made from silk I've kept around for years. There's nothing quite like silk on a hot summer evening."

JH showed us into the living room, which overlooked the bay.

Billy soon reappeared wearing a starched white apron, with "Guatemala" spelled out in bright embroidery on the belt and pockets. She carried a small oval platter and asked if we'd like to try her latest "culinary creations. An amuse-bouche, anyone?"

There were two choices: one small, turquoise-blue brie cheese with black currants melted onto round rice crackers; and the other, pinwheel sandwiches of smoked salmon, crushed pineapple cream cheese rolled around pickled asparagus, and thin slices of celery.

They were delicious.

"Such a hot summer night," Billy said, "more drinks before dinner?"

We had a choice of ginger beer with lemon ice or home- bottled sarsaparilla. I chose the ginger beer since we bottled our root beer.

My Freddie, my older brother, wisely chose the root beer, while my younger brother, John, said, "Both. I'll mix them." As the adults settled into their pre-dinner conversation, I jumped out of my chair and ran to Billy.

"Chilli, curry, hot! Hot! Chutney!

Tears welled in my eyes. In a flash, we were in the kitchen, Billy pouring lemonade over ice, and me screaming, "Chilli! Chutney, curry chutney! That's what I need!"

Billy handed me the glass of chilled lemonade and asked, "Chutney? Ah, chutney to go with the hot curry. You've eaten hot curry?

I nodded, "Oh yes! In Fiji."

"Then you must see my chutney preserves, because we won't be having any with dinner. How's the lemonade?"

"It's fine. I'm fine now What are we having for dinner?"

"That will be a wonderful surprise," she replied. "Now let me show you my chutney resources."

In a decoratively carved wooden tray, "from Srinagar, in Kashmir," she said with pride, eight glass jars of varying sizes nestled neatly in their grooves. Each one caught the light differently, hinting at spices, salts, or preserves within. From a nearby drawer, she retrieved a slender box lined with eight delicate spoons, each just the right size to match the jars.

"You may taste a level coffee-spoon amount from one jar, your choice. Do you think your brothers also want a taste of chutney before dinner?"

"I don't think so. They're worried enough that you're going to serve steak-and-kidney pie."

"Why in the world would they think that?"

"Because you're English, I guess."

Billy laughed. She filled my glass with more lemonade and led me to a kitchen table by the dining-room door. The closer I got, the more difficult it was for her to keep from laughing.

There, on a polished silver warming tray, was a large, golden-crusted pie. She pushed the dining-room door open and held onto both sides of the door jamb to keep from falling over laughing.

I had to ask, "Steak-and-kidney pie?"

"No," she said. "Chicken. But you must play along with me when we seat your brothers."

I nodded my agreement.

"Now look at the rest of the things we're going to eat."

There were six glass plates with pink and white molded fish shapes surrounded by radishes and cherry tomatoes. In the freezer, there were glass plates with mint leaves and small scoops of what appeared to be yellow popsicles. After we'd seen everything, Billy asked how we call people to dinner at my house on special occasions.

I could have told her we rang the dinner bell from the P&O–Orient Lines *SS Orsova*, but wanting to show her I knew how to speak Māori, I said, "Kai Kai!"

"Oh well, then we'll go out and invite everyone to, how do you say it again."

"Kai Kai."

I would like to say we danced into the living room as if performing a two-person conga line, but Billy was a reserved and refined woman. When we reached the living room, she took off her apron, and we both said, with dignity, "Kai Kai."

While we were seated around the enormous table, Billy came to me and said, "Remember our surprise!" Then she brought out the jellied-fish dish and told us it was Coho salmon, crab, and halibut.

We had been warned to use our manners, especially not to eat anything until grace was said. My brothers and I stayed composed until JH said, "Shall we say grace?"

"Sure, I know one!" I said.

This was not the most unmannerly response, but JH wasn't expecting it since he already had his long, tanned fingers poised.

"Would you like to say your grace, Jeremy?"

My brother, Freddie, had already given me a side kick in the hip, and I said, "Oh, no, yours will be fine, thank you."

Even the kick couldn't dispel my growing sense of anticipation around whatever embarrassment might surround my brothers over their steak-and-kidney pie aversions. We bowed our heads, and JH began.

For the sun and for the rain, for the fields and for the grain, for those who bring the fish to us, for those who labour for the harvest, we give thanks and praise your name as we celebrate a newborn babe.

I had never heard a grace quite like that, and I'd also never had such beautiful food, delicious and tastefully served. Billy and Dad talked as we unfolded large linen napkins. JH made a point of smiling at us as he chose a small fork and what looked like a butter knife and started to eat.

With every bite, he smiled to himself as he savored the taste, and still, not speaking, smiled at us with his eyes. The jellied- fish dish was the most wonderful food I'd ever eaten.

When my younger brother started making faces over his second spoonful, I went over and took his plate, put a glass of water in front of him and whispered, "Wait until you see the pie!"

I thought the big moment would come right after the fish plates were taken away, but what came next were the icy scoops on two different kinds of mint leaves: peppermint and pineapple-flavored. My dad looked at us, thinking this was dessert, and he smiled.

My brothers looked at one another and didn't say a word; I was sure the same thing was uppermost in their minds: we eat this dessert, then we have steak and kidney pie for an after-dessert. Both looked disoriented.

Billy caught their attention. "Sorbet, anyone?"

We all said, "Sure!" and reached for the plates. Billy explained that the sorbet was made from lemons, apples, and a few very tart crab apples.

"It's to clean your palate before the other tastes of the meal, new green beans, asparagus, the meat dish, the pie!"

She smiled at me and winked, raising her eyebrows.

Both of my brothers made horrible faces as they bit into the sorbet, but I tried to ignore them. JH discussed gardening and farming with my dad. He knew a lot about everything. But I knew another seed had to be planted and, now and then, I turned to my brothers and whispered, "Pie, pie…"

Billy took away the sorbet almost as soon as we started eating it. I finished mine, Freddie had picked at his, and, after the first few slurps, John had stuck a fork in his, tearing the mint leaves and chewing them before spitting them onto his plate.

Billy said he didn't have to eat them if he didn't want to. "They're just garnish, decoration."

By this time, I felt certain both my brothers thought that, although Billy was the head dietitian at the local hospital, she was the strangest cook either of them had ever encountered. They had no idea what they would be forced to eat next.

The mystery was solved when she brought in the pie, this time, on an even bigger silver tray, with cooked beans and browned potato halves encircling it. She went back to the kitchen and, this time, reappeared with two silver chafing dishes.

"I had to go to the cold room's coldest spot for these, but they've been on ice since five. And listen!" She shook both dishes, and they jingled. "There are still ice pieces."

While my brothers anticipated God knows what after being primed with the jellied-fish dish and the sorbet, I examined the two silver trays of chilled asparagus. They had small silver fork spears balanced on hooks soldered to the edges of the dishes. *What treasures*, I thought.

"I hope you all like a meat pie."

I looked at my smiling father first, then at both my brothers.

They were more than ready to ask to be excused. "Chicken pot pie, anyone?"

"Whew!" Freddie said. "I thought it was going to be

something else."

"You don't like chicken pot pie, Freddie?"

"Oh, I like it very much. I'll probably have two helpings."

I turned to him. "Pig! We haven't seen whether there's enough to go around, and you're already asking for a second helping."

"I'm sure there'll be enough to go around," Billy said. "May I help you to some greens, Freddie?"

By this time, my younger brother discovered it wasn't steak-and-kidney pie, reached for a plate, and pushed it across the table with two forks.

"I just won't have any of the chicken's guts, if you don't mind."

Everyone thought this was funny and set into a good meal.

We were allowed to eat the asparagus with our fingers and heap our side plates with French-bread dinner rolls and butter. The butter was served in two silver dishes, one with a rounded top ("like a dome in the Kremlin," JH said) and the other with a rectangular lid.

"Bauhaus," Billy said, pointing to the butter dish at her end of the table. "It was a modern school informed by the Arts and Crafts movement and Art Nouveau."

"Is Bauhaus a school like Friedrich Froebel's school?" I asked.

"No, nothing like it. Bauhaus was a school of art and design, and it was remarkably modern for its time. But why not do a Froebel experiential learning exercise?"

I knew from my grandmother that Friedrich Fröbel had founded the kindergarten movement, but I had no idea what Billy meant and worried she would think we were stupid.

She invited my brothers and me to take apart the butter dishes. Both had small glass saucers that sat on removable silver trays, underneath which there was ice to chill the cubes of butter.

"But if either of you would like some ice, feel free to pass it on a spoon."

I took the ice and carried it back to my seat in an egg-and-spoon style. At every setting, there was a small bowl with floating lemon and lime rings and tiny pink globs at the bottom.

"Sundried grenadine-and-rosewater flakes," Billy said, "my invention."

She knew we didn't think these bowls looked appetizing. "Finger bowls, boys."

After my brother had eaten all the chicken pot pie and vegetables he could, he discarded some of the unwanted food into his finger bowl and washed his fingers in it. Then, he reached for his dessert spoon. There began a scooping out of the liquid as he bent forward and slurped it off his spoon as though it were cold soup.

Until we stopped him, he plodded away with serious effort etched on his face. He didn't pretend to enjoy the meal again until we got to the dessert, which consisted of eclairs with lemon curd custard, and salmon berries, along with strawberries, raspberries, blueberries, and whipped cream.

After dinner, we moved to the living room again, and Dad and JH talked about the Old Age and Canada Pension Plans "as an income supplement for contributors and their dependents."

JH believed the universal Old Age Pension was a noble idea, something that dignified all people in their later years. Still, he often said, "I couldn't, in conscience, cash the cheques…" The words hung in the air, half conviction, half confession.

Without another word, he stepped into his study. Moments later, he returned, carrying a long, polished metal box, weighty and faded, with a crest embossed on its lid. Whatever it held, it had clearly been waiting for just the right moment.

"I keep them in here, in a deed box, but I never cash them.

How could I?"

I knew JH and Billy were wealthy, and I knew they were good people, but I couldn't imagine someone getting an Old Age Pension cheque each month and not cashing it.

"You could cash them and give away the money," I said. "Aha!" JH replied. "We have a discourse!"

"What's a discourse?"

"It's an exchange of ideas on subjects of importance. Take, for instance, your suggestion, Jeremy, that I cash the cheques and give away the money. I have several reasons why I don't wish to do this. If I cashed the cheques, I would have to add their value to my other earnings. I would then be taxed on the added value in my bank account. And to whom should I give the money?"

"To poor people."

"And why are there poor people?"

"Lots of reasons."

We were invited to discuss various forms of poverty. Dad joined the discussion.

"Some people are not adequately paid for their work, and then there is unemployment."

JH listened carefully and questioned Dad on "scabs and people who don't stand with the union. This undercuts the collective interests of the labour movement and creates conflict in communities among other working people. But, as regrettable as these impacts of poverty may well be."

Billy interrupted. "As regrettable as these impacts of poverty *are*, my dear."

"Yes, Billy, quite correct. All forms of poverty are regrettable, but it is important to see forms of poverty as symptoms of the greater problem, not the central problem itself."

I marveled at how well he could speak and wondered when I should mention that a few bucks could surely help buy toys or books for poor kids, but I didn't find the chance.

"Does anyone agree it's likely we have poverty because the resources, the wealth of the land are not properly distributed among all people?

"Of course," my father agreed.

"Then you must as well agree it is the responsibility of government to ensure the wealth of the world, particularly in a rich country like Canada, should be distributed fairly among all her peoples."

This sounded reasonable, and we all agreed.

"But there is poverty," he said. "In this community, even though we've embraced the establishment of significant heavy industrial production."

Dad asked whether JH believed pensions should be reserved for those with low incomes and, if so, how eligibility should be determined.

JH leaned back thoughtfully. "That's a very good question," he began. "In my view, everyone who reaches a certain age deserves the opportunity and the dignity to step away from the workforce. And under no circumstances should a retired person be subjected to a means test to qualify for that support.

"If you create a system where only one segment of the elders receives the pension, you risk dividing the elderly into classes. That leads to quiet but corrosive consequences, a social stain, however subtle, on those who rely on the pension as their sole source of income.

"Of course, those in real need may not be discouraged by such stigma. But what about those on the borderline? Or those who, for any number of reasons, feel compelled to keep up appearances?

They'd face invisible pressures, not to apply, not to admit they need help. And in doing so, they might forgo the very support that could sustain a self-reliant, decent life." We politely listened.

"I may choose not to cash my pension cheques, but I do so mostly for my reasons. It would be an *indulgence*. At the same time, I staunchly affirm pensions should be available for all who choose to do with them what they will."

"But that doesn't answer why you shouldn't cash them and give the money to the poor, or the church, or s*omebody*," I asked.

"I think it does. If my charity, from time to time, helps someone out, how can I be certain that the little I give will adequately assist that person? Perhaps my generosity would only perpetuate circumstances that, for a family or an individual, should be altered *adequately* by full-time employment or some form of state-subsidized training.

"All that is good, too," my father added, and JH went on.

"I hold the belief that I've liberated myself through my fortunate circumstances from the type of society inherited since the last war. I can do one thing today and another thing tomorrow. I may wish to hunt in the morning, fish or tend my garden in the afternoon, and critically discourse in the evening without becoming a hunter, fisher, gardener, or critic. And, especially, in a land like Canada, I may enjoy these spheres of active engagement at this stage of my life.

Billy had her say. "When I'm eligible, however, I intend to cash my pension cheques because I believe I've earned this acknowledgement of my status as a worker in healthcare, a fundamentally necessary institution of civil society.

"I hold the view that my pension should be spent entirely on those things that will enhance my enjoyment and well-being. If I spend wisely, I'll be creating employment for those who produce and distribute high-quality works. Fred, you appreciate how sometimes an engine needs to be primed.

When I retire, I intend to revitalize the economies of floral nurseries and art societies, and in the process, cultivate various bouquets. Perhaps even some interesting watercolours or oils, or even particularly thought-provoking abstract art."

Billy steered us to a discussion of art and showed us framed paintings hung throughout her home. After we toured stately, colourful paintings, the hour was late.

"On behalf of all of us," Dad said, "I want to thank you for your hospitality tonight and especially for the good conversations. We've not only enjoyed the wholesome food so beautifully served but the food for thought as well."

As we climbed into the car, each of us was handed a small, personal card of congratulations. Printed on every one was a child's grace, a thoughtful gesture that felt at once playful and profound.

Mine read:

### *"Rub a dub-dub. Let's have the grub. Yea! God!!"*

It made me laugh, the rhyme sharp and strange in its cheer. John's was more lyrical:

*"Thank you for the world so sweet, Thank you for the food we eat.*

*Thank you for the birds that sing, Thank you, God, for everything!"*

Freddie's bore the unmistakable tone of Robert Louis Stevenson:

*"It is very nice to think.*

*The world is full of meat and drink, With little children saying grace*

*In every Christian kind of place."*

Mom and Dad shared a more reflective verse:

*"Creator God, let us be thankful.*

*For whatever light, laughter, food, and affection may come our way.*

*And let us be mindful equally of those who, At this or some future moment,*

*May be sadly without any or all these good and glorious things."*

The car hummed gently as we rode along, each of us quietly holding our card, touched in our own way.

# CHAPTER 15

**Bumholing and Blackballing**

If getting two front teeth knocked out and several years of conflict with the school principal was a Camelot of innocence, that relative state of grace, and recollections of it, vanished by the time I became an air cadet.

The worst things that could happen to an air cadet were being "bumholed" or blackballed. Neither ever happened to me, but throughout my three years as a cadet, there was the ever-present threat.

We enjoyed weekends and summer camps on air force bases, and we sang on bus trips.

The washroom walls are soaped up,

There is water in the halls,

And without the Flight's permission,

His initials are on the walls.

And when a Sergeant bugs us,

We smear polish on his balls

As we go marching on.

We also sang "The North Atlantic Squadron."

We were on the good ship Venus

Oh, boys, you should have seen us

The figurehead was a whore in bed

And the mast was a ninety-foot penis.

A way hay and a fife and drum

Here we come, full of rum

Looking for women to peddle their bum

To the North Atlantic squadron

The cabin boy, the cabin boy

The dirty little nipper

We filled his ass with broken glass

And circumcised the skipper.

A constellation of historic and cultural issues structures and creates a sexual status hierarchy within military and paramilitary organizations. Sexual harassment, exploitation, and oppression are elements that define that hierarchy.

During times of armed conflict, small boys worked as powder monkeys, carrying the fuses and flint matches between the guns on the great man-of-war vessels. They often burned their hands replacing the used fuses, essentially cotton rags sprinkled with gunpowder.

The burns rarely healed properly, were often infected, and, if the powder monkeys survived to adulthood, their scars still throbbed, serving as indicators of their class origins.

Historic naval superstitions proscribed the inclusion of women aboard ship. It is reasonable to assume, at the time of boarding a ship, that an assessment of a child's gender took place. The degree to which such an assessment was sexually abusive warrants consideration. What happened to the girls found on board the ship? They were likely thrown overboard. What was life like aboard ship for the boys?

When we sang about the cabin boy in The North Atlantic Squadron, the song told about a child who had been sexually brutalized. But we were adolescent boys in blue serge uniforms, Windsor knots at the necks of our ties, and white-braided lanyards looped through our epaulettes. A busload of boys singing was sufficiently removed in time and geography from the sadistic rites of pre-colonial seafaring culture. Or was it?

The sexual violence of the skipper's public circumcision was abstracted by the social context in which we were participants. Riding on a bus, paid for by the military, we went for a weekend camp at a military base.

We rode in an aircraft, never a plane, as we were firmly instructed to say "the aircraft." We ate in the mess hall and slept in a shared dormitory, the days folding into one another with military efficiency. Curiously, we punctuated our singing and occasional storytelling with the phrase: *"War is hell."* No one ever explained why. It just slipped into our speech like a refrain, part joke, part truth, part borrowed wisdom from a world we only half understood.

From the back of the bus, a murmur emerged, a ripple of snickers. The murmur was whispered and snickered about, and everyone wanted to get in on what was being said. The youngest or the newest were deliberately excluded from the whispered messages. When the bus was divided into camps of those who knew and those who didn't dare ask, one of the youngest members of the group, who knew what had been discussed, approached some of those who didn't know and said, "Blackball tonight."

This part of the excursion always terrified me. As the bus trip proceeded and more stories were told, others from the group in the know approached those who were not in the know. They winked or pinched and approached to just laugh in someone's face. If we stopped to get something to eat or go to a washroom, two or three of the older cadets approached, looked fierce, and whispered, again, "Blackball tonight!"

At the pit stops, there were always power games. The oldest cadets went into the gas station washrooms and stayed there. Some emerged from the washrooms to taunt designated victims.

"You haven't gone to the can as yet!"

"We'll make fun of you for years if you wet yourself before we get to the base."

Hey, you should go in there; they're bumholing Timmie; you should get in on it."

It was usually the cadet who had been the initial messenger to the victim group who emerged, swearing, from the washroom, his tunic open, his shirt soaked with water or urine. He punched anyone waiting and too frightened to go in.

"They didn't do anything to me," he vehemently exclaimed, directing his glare at the closest cadet, "but tonight, we're going to bumhole *you*. Then we're going to blackball *all* of you!"

Neither of these activities was associated with anything pleasurable. The threats were manufactured by highly stylized behaviours on the part of the senior-ranking cadets and their underling collaborators.

A typical pit stop concluded with a call to get back on the bus. Most of the older cadets took their time as they left the washroom; only one or two remained. A corporal or a cadet sergeant ordered the youngest cadets into the washroom.

"So, as you don't piss yourself on the bus. Now get in there on the double!"

We got in there and fumbled with the buttons on the flies of our pants. This usually occurred while the most senior cadet, often a warrant officer, occupied the urinal. Then he turned around, exposed his penis, and shook off the drips. Sometimes he peed on a younger cadet's shoes or pants.

But we didn't hurry; he took his time to put his penis back in his pants, doing up each button with some ceremony. When he had finished, he bent his knees and jiggled his hips to arrange his testicles appropriately.

During all this activity, the process by which someone would be blackballed was explained. His quiet, controlled voice was a form of sexual terrorism. The youngest cadets dared not speak in case they missed a casually mentioned name.

Sometimes, when the bus horn sounded, we would push past him to use the restroom, and he would allow us to approach a urinal. Or another of the older cadets moved to allow a younger one into a cubicle. The blackball lecture went like this:

"If you find your shoe polish missing from your kit when you get there, don't say anything. You can look for it. It might be in a furnace room or by a heater, or by a window where the sun has been shining. Most likely, it will be on or near a heater. If you want to save your nuts, you won't say anything."

"When the adult officers go to the mess tonight, someone is going to get blackballed. Best if we have a volunteer. Be like Timmie, he volunteered. Now he's one of us... Well, most of the time."

"If we don't get a volunteer from one of you, we'll have to wait until the adult officers go off to a party at the PMQs (permanent married quarters). That's when it will happen. Perhaps we'll wait until they return. They'll be pissed drunk and in another barracks, and they won't hear a thing. But if we wait, it'll be harder to

blackball you."

He would catch someone's eye, stare at them, and repeat, "*You.*" Then he'd lock his gaze on another young cadet. "*You.*"

A deliberate silence hung in the air. Then he spoke, his voice low and measured, too calm for the words that followed.

"The shoe polish might be cold. That's why we wait until you're in bed. Then eight of us jump on you. If you lie still, it won't be so bad, just the sheet yanked off and a pillow pressed over your head. We lift it now and then so you can breathe, but only if you promise not to scream."

"Two hold your arms. Someone holds your legs. One guy lies across your chest. Whoever's left? He grabs your penis, if he can find it. If you've got one. And then everyone else smears polish on your balls. That's the whole thing. No big deal."

He paused, letting the tension breathe. "But if the polish is cold? That's when it stings. And it'll hurt more when they poke you." He looked around. "Any volunteers?" No one spoke.

"If any of you say a word about this, you lose a nut. Then the other. Most keep quiet. But I'm worried about you, Jeremy. You're a talker."

The room seemed to shrink.

"If the blackball shits himself, it's easier. Makes it simple for him to get taken advantage of by everyone. You've seen what we use. I saw you staring when I was doing up my fly. Don't let that happen again. Eyes forward."

He might straighten his tie or look in the mirror before ordering us back onto the bus.

No one I knew was ever blackballed or bumholed. I was always terrified it would happen to me, and I don't know why I didn't quit cadets.

It would have been impossible to tell an adult about any of this. If anyone squealed, it was perceived that they would be dead. And it would be a horrible death.

We took plane rides and sometimes even took control of the glider. We also shot rifles at the range. On the runways and at the range, we were rigorously disciplined.

"No tomfoolery around here, or back to the barracks for you, and that goes for anyone! Absolutely no tomfoolery. Does everyone understand?"

"Yes, sir!"

The adult officer in charge stared us down until the silence became oppressive. Older cadets, the ones from the back of the bus, squinted into the sun glaring on the runway, or into the brightly lit targets we were to shoot at. Sometimes, they turned their heads, squinted harder, and clenched their teeth, flexing the muscles along the sides of their faces.

The squinting, staring, and flexing jaw muscles were an elaborately ritualized form of collective behaviour that reinforced a conspiracy of emotional terrorism. From the perspective of graduate school, the behaviour degraded us all.

Those who made the threats, and some of us who received them, knew that threats of sexual violence were effective ways of maintaining control. If the adult officers knew anything about what was going on, they didn't deal with it in any effective manner.

The occasional exhortation to avoid tomfoolery did nothing to diminish the terror I experienced whenever no adult member of the squadron was present. Maybe tomfoolery was something else, an equally abhorrent activity I was lucky never to become intimately familiar with.

# CHAPTER 16

**Summer Garden**

When I was in junior high, I wanted to quit my paper route for numerous reasons. I was bullied by two big, ugly high school boys, and there was no way out but to quit. I saw Billy and JH at the Easter service, and they asked how I was doing.

"I'm really fed up with doing my paper route and going to school. I just want to take off and become a film star."

"What would you do for money while you were auditioning for parts?"

"Oh, I don't know. I don't like washing dishes, and I don't want to be in the newspaper business all my life."

Billy laughed, but JH said, "It's not as if we all may be rulers of the King's Navy, is it?"

"Certainly not with the Victoria Colonist," Billy said.

Easter Monday was a holiday, and JH came around to our house to show me how to layer hydrangeas. He asked if I would like to come to his house later and talk about summer employment as an undergardener.

When he left, I asked Mom if I could have a summer job.

"Sure. Your dad's local may go on strike to break away from the international union, so we're not going anywhere."

I took off on my bike. JH met me in the driveway and showed me around the yard. He said I didn't need to do anything other than water the various gardens and whatever else suited my interest.

He showed me the garden tools and how to take care of them. If I could attend the next three Wednesday afternoons after school, I would have a good sense of what needs to be done by the time they leave.

"The garden will be your own private Xanadu from mid- May until mid-August."

There was only one rule: under no circumstances was I to invite other children, except my brothers, to help in the garden or to swim at the beach.

I was not to encourage Jake-Jack-John Dudley to come anywhere near the property. I told him Jake-Jack-John was in a foster home somewhere, and I didn't expect to see him until I was at least forty.

I returned to the estate one afternoon and found JH in the garden shed sharpening clippers. I knew he'd wiped his eyes with his gloved hands, as there was a smear of oil under each of his eyes.

"JH, have you been crying?"

"Why would you ask?"

"It looks like you wiped tears from your face, not just your eyes, but tears that ran down your face."

"Well, no. Yes. You see, Mrs. Whittaker-Ramirez and I are going to Europe for the summer."

"You and Mrs. Whittaker-Ramirez? Are you going to Montparnasse, where she met Ramirez? I thought you and Billy were going to Europe."

"We are. Isn't that what I said?"

"No. You said you and Mrs. Whittaker-Ramirez were going to Europe."

"I did? I must have. No, Mrs. Whittaker-Ramirez is moving to Victoria. Billy and I are going to Europe."

He began to sniffle, then sob, "You must excuse me. Forgive me?"

"Nothing to forgive. You're just sad because Mrs. Whittaker-Ramirez is moving away."

"Yes, that's right. She was such a good chum and wise in the ways of the world. I'll miss her. She made some money from one of her inventions, and she's going to Victoria, as she believes there will be more opportunities for Eileen Kathleen."

"I'm sorry, both are leaving," I said. "But we can go to visit them in Victoria, maybe . . ."

"Maybe," he replied, "but first, let me show you how to deadhead the roses. I must show you how to identify deadwood, how to cut it back to the lowest bud to generate new growth. And we must talk about the working day and the surplus value of your labour to me, your employer."

I sensed JH was telling me he was the boss, but he was fair from then on. He told me that even if I wore gloves, I might prick my skin on a thorn, and that would be my general oppression as a worker. Furthermore, all workers were generally exploited by their bosses, but as I was his only employee, he wanted to ensure that I experienced as little exploitation as possible.

I would receive ten dollars a week for whatever chores I wanted to do, but, in particular, I was to cut the grass and ensure the roses didn't overgrow.

I was also to "enjoy the surplus value" of my labour. This meant I was permitted to pick the finest roses for my pleasure, and I was not to let any of the responsibilities of the garden overwhelm me. I was to take time to read and drink as much Marks and Spencer orange cordial as I liked.

I could also tend a small plot of my own and grow whatever I liked. That would be another aspect of "the surplus-value" of my labour that your employer would not appropriate." JH called these instructions "the practice of the German ideology."

Before he and Billy left for Europe, JH gave me an envelope of postdated cheques totaling one hundred sixty dollars. He shocked my parents, not only with his confidence in me, but because they heard him tell me, "In whatever you do, practice what you've learned of the German ideology. Garden and hunt and read and memorize without seeing yourself as merely a gardener, a reader, or an actor."

What could my parents do? JH and Billy were driving away.

Lydia was at the house when they came to give me the cheques and, after they left, she said, "Vas is zis? Zay asked him to practice ze German ideology?"

"It's Lenin," my mother replied. "The German Ideology is for Marxists what the Sermon on the Mount is for Christians."

"It's nothing of ze kind," Lydia asserted. "As Winston Churchill said on the BBC in 1943 of za Huns, 'It's either them being at your throat or your feet.'"

Mom and Dad tolerated her remarks about Germans because they had arrested her. However, since the war was over, they occasionally expressed views that did not blame German working people for decisions made by the Nazi state.

"You've taken zehr money, now you must learn zehr anthem." With that, Lydia shocked us when she began singing.

"Hitler, he only had one ball,

Goebbels had two, but they were small.

Goering he sure was boring, and then, that Himmler had no balls at all!"

The room fell briefly silent after Lydia's spirited recitation.

My mother offered a quick, half-apologetic smile. "Lydia's vulgarity is harmless," she said, as if reading the air. "It's just an old propaganda song, from the war."

Yes," my father agreed, "but all humor has some qualities of truth when someone uses it for propaganda or satire."

Both agreed with Lydia that Billie would be all right. Still, it would be a shock when JH found out that Germany richly benefited from the Marshall Plan and the infusion of capital to build a modern, capitalist economy.

I told my parents that sometimes JH got confused, and once had told me that he and Mrs. Whittaker-Ramirez were going to Europe. (Likely, I thought, *he and Billy were going now so that he'd see the place again before he lost his mind and ended up consorting with institutionalized people like Mrs. Hennie).*

I was instructed never to discuss anything I may overhear around JH and Billy's house. That was an assumed element of the confidence they had shown in me, never discussing what I heard was called "probity and discretion."

One day, while I was working around the garden, their nephew, Charles, the photographer and Cub master, drove up and criticized everything I was doing. He said I wasn't allowed to pick the flowers, particularly the roses. They weren't ready and, anyway, I'd been picking columbines all summer. I'd gone out of my way to water the hydrangea transplants on the path to the beach. I felt so upset I couldn't tell him off. What really annoyed me was that he didn't even remember my name and kept calling me "boy." Or saying, "Listen here, boy."

I never told my parents he had acted like a pompous ass.

Before JH and Billy left, JH explained that he would create another garden plot by placing a large, fourteen-by-twenty-foot tent on a part of the lawn.

Inside the tent, he and Billy arranged a large Indian rug, a desk, a carved Kashmiri modesty screen, a single bed and mattress with heavy cotton sheets, and a couple of pillows. There was also an old writing desk with a small bookshelf beside it.

Along the wall, above the desk, pinned to the ceiling of the tent, was a map of the world with various places marked "Near Brussels, Waterloo, 1815; Dunkirk, June 1944; Bayeux, 1066; and, in red pen, where they would be in Europe on which dates.

An envelope lay on the desk. It was an invitation for me to stay in the tent, and it stressed that if the weather was especially hot, I should ensure the gardens were watered at night and early in the morning. On a special sheet of paper was a handwritten copy of "Xanadu." By Samuel Taylor Coleridge.

*"In Xanadu did Kubla Khan*

*A stately pleasure-dome decree:*

*Where Alph, the sacred river, ran*

*Through caverns measureless to man*

*Down to a sunless sea.*

*So twice five miles of fertile ground*

*With walls and towers were girdled round;*

*And there were gardens bright with sinuous rills*

*Where blossom'd many an incense-bearing tree;*

*And here were forests ancient as the hills,*

*Enfolding sunny spots of greenery."*[6]

On the back, Shelly's poem, "Ozymandias."

*I met a traveller from an antique land,*

*Who said— "Two vast and trunkless legs of stone*

*Stand in the desert... Near them, on the sand,*

*Half sunk, a shattered visage lies, whose frown,*

*And wrinkled lip, and sneer of cold command*

*Tell that its sculptor well those passions read*

*Which yet survive, stamped on these lifeless things,*

*The hand that mocked them, and the heart that fed;*

*And on the pedestal, these words appear:*

*'My name is Ozymandias, King of Kings;*

*Look on my Works, ye Mighty, and despair!'*

*Nothing beside remains. Round the decay*

*Of that colossal wreck, boundless and bare*

*The lone and level sands stretch far away."[7]*

On a summer afternoon, while I worked in the garden, the bank manager and the art teacher, accompanied by bongo drums, drove up to the property in their flashy red sports car.

Billy had invited them to use the place while she and JH were away. They immediately recognized me, and I served them orange cordial from the cold room.

Politely, respectfully, and very formally, they showed me a letter of introduction addressed to me, encouraging me to "show the bearers where to find the house keys and show them the hospitality of the estate."

We had a delightful visit. They brought various ethnic foods from Vancouver, including tins of Greek dolmades, jars of olives and smoked salmon, Belgian endives, and several different kinds of cheese. The teacher had returned to university and became an art education professor.

Although I showed them where they could sleep, they asked if they could use the rug in the tent. I agreed and went home while they were there, but they first asked me to get my parents' permission to take me to Victoria's Butchart Gardens.

Of course, my parents said that was fine, and I loved riding in the back of the sports car while with my two most successful adult friends. I loved them, and I think they loved me.

After the Gardens, we went into Victoria and had dinner in the Bengal Room of the Empress Hotel. The Indian food was spectacular, and the servers were extremely respectful. While we ordered, one asked, "Would your son like a Shirley Temple?" We all laughed heartily.

"No," I said, "I'm more of a Judy Garland type kid. Any drink named after her?"

During dinner, we discussed what it was like to be me, my joys, my struggles, how I loved being the undergardener at JH and Billy's, how I hated Air Cadets, and why. They were shocked, as it was abusive, and both Tristan and Adrian told me I should quit.

I explained my older brother had applied for a Regular Officer Training Plan scholarship, and it wouldn't look good if his younger brother had dropped out. Having me for a sibling would be a stain on my brother's character. Besides, I liked playing "God Save the Queen" and "O Canada" on the piano at annual inspections.

They stressed to me that if it became too much, irrespective of my brother's scholarship application, I always had a choice. I may have a few other choices that are equally important in my life at this time, but quitting could make me feel like I had done something positive.

I don't know why I never kept in touch with these beautiful, successful, intelligent men. Even knowing they existed made my life bearable and allowed me to be me with a quality of style and wit.

# CHAPTER 17

**The Fussy Hussies from Tussie**

Crofton kids had to take the bus to the junior secondary school. I lived on the hill, attended the Anglican Church, and took catechism class. Every weekend, I babysat my neighbours' kids and usually had more spending money than the other kids. For whatever other reasons, I was also bullied in numerous ways.

They called me Oberon, King of the Fairies, after I played the role in *A Midsummer Night's Dream*. But some kids also called me Organ Grinder, because of a chipped front tooth I hadn't yet learned to hide when I smiled.

One afternoon, a couple of tough guys stole my lunch and made me buy it back for two dollars. After that, I stopped taking the usual bus. I took two others instead, longer, out of the way, and that's how I fell in with the Penelakut girls from the Tussie Reserve.

They called themselves the "Fussy Hussies from Tussie," and they went to the Catholic school. They were sharp, funny, and surprisingly kind to a kid like me.

We got along well. They knew about the Kanaka Hawaiians, who had escaped from the Hudson's Bay Company fort at Fort Langley, and thought it was cool that I'd been in New Zealand and knew Māori songs.

When elections were announced for student council, I was

nominated to run as president on a platform of more lunch-hour sock hops and four school dances each year. We then had numerous lunch-hour sock hops in the gym, as well as three Friday-night dances, a pre-Christmas dance, a spring dance, and an end-of-term dance.

After the pre-Christmas dance, there was considerable enthusiasm for entertainment at the spring dance. I had arranged with the home economics teacher to make different kinds of finger food, and a punch made with dried mint, pineapple juice, ginger ale, and peppermint tea, along with various citrus fruits frozen into large ice cubes. With paper cups and colored paper napkins, it was a festive and sophisticated event for a junior high school dance.

Sometime between Valentine's Day and Easter break, I negotiated with the jocks to participate in a half-time drag show. All the hottest males would dress up like women, if, during the last dance, while "House of the Rising Sun" played, I went up on the stage and turned off all the gym lights.

To ensure we pulled it off, we agreed that instead of doing this when the dance was supposed to end (at 10:30), I would turn off the lights at 10:15 while the chaperone teachers cleaned up the home economics room.

The drag show went surprisingly well. Along with ballgowns, nurse, and secretary outfits, my friend, Sean, whose mother was a social worker, lent him pedal pushers and fancy sandals. He wore a Madras-print blouse tied below his ribs to show his bare midriff, and the outfit was highlighted with two big balloons for breasts.

I dressed in a slinky flapper shift with strands of plastic bugle beads. When I came out swinging the beads, one of the navy cadets played stripper music on his trumpet.

After the entertainment, we resumed dancing. Most of us who had participated in the drag show continued to wear our costumes. When the first notes of "House of the Rising Sun" began, I locked the back door to the stage.

The speakers and other dance paraphernalia were on the stairs, and the curtain was pulled. When the lights dimmed, I turned up the music and honored the deal I'd made.

The principal at that time was Grant Garnett, a generally reasonable man. His red hair was turning grey, and he wore it in a flat-top crew cut. After rattling the stage door but having no success in summoning me to open it, he ran back to his office for a set of keys.

By the time he got back, the dance was over. The curtains were open, the lights were on, and I mingled with other students on the dance floor, accepting their congratulations for pulling off the blackout plan.

Mr. Garnett, nicknamed Scar Balls, thanks to his crew cut and a persistent rumor that he shaved his pubic hair and sometimes nicked his own scrotum, found me quickly. He grabbed me first by the ear, then by the shoulder, and frog-marched me down the hall to his office.

He was livid.

"We will only discuss your discipline in the office without any witnesses," he said.

I was scared.

As we neared his office, we saw, down the hall from the main entrance, Billy, her cousin, Phyllis, their cousin, Kathleen Long, and Barbara Thornton-Sharp, the school's art teacher. They had dropped in on their way home from the Victoria Symphony concert at the secondary school.

Billy and Phyllis sensed there was some *frisson* between Garnett and me. They immediately rushed over and pulled off a coup de grâce that was a remarkable expression of their social savvy, class consciousness, and moxie.

"Jeremy!" Billy said, "Thank God we're here in time to see some of the fancy-dress costumes. And congratulations on being selected Head Boy."

Garnett looked at me with a malevolent sneer.

"We don't use that term in Canadian schools," he said. "Jeremy is merely the president of the student council. Head boy responsibilities usually involve an enforced hierarchy that involves fagging. And we haven't had anything like that sort of thing in this school until tonight."

He glared at me, still holding onto my bare arm as I stood beside him in my flapper dress.

"Mr. Garnett," Kathleen said, cutting in, "that is beneath you."

Billy added, "Grant, you obviously know nothing about the practices of fagging."

Phyllis approached and took my free arm.

"Grant, can you not let Jeremy escort me to the auditorium so I may see some of the other students in their fancy-dress costumes? "

Somehow, Billie disengaged Garnett, and Phyllis and I hurried down the hallway. Before entering the gymnasium, she took off her Kashmir shawl, arranged it sash-like from her shoulder to her hip, removed the hulking diamond broach from her dress, pinning it to her shawl, and tied the shawl at her hip. Billy soon caught up with us.

Both women were dressed well. Billy carried a white fox fur stole; both wore diamond necklaces, earrings, and bracelets. Phyllis was dressed in dark green silk brocade, while Billy wore a modern, elegant raspberry party dress featuring a black-lace overskirt with sequins.

In the gymnasium, students pulled down streamers, supervised by teacher-chaperones.

I went to the microphone.

"Fellow students, may I have your attention? Miss Thornton-Sharp's stepmother and her cousin would very much like to inspect those still in costume and anyone else who would like to greet them."

Everyone caught on to my stalling, an obvious ploy to avoid Garnett and began forming a line. First came the boys still in costume, then the couples, arms linked and eyes scanning the crowd.

Last were the girls, each having spent time and no small amount of money, getting dressed and coiffed for the occasion. No one wanted the moment to be wasted, least of all them. "How do you do?" most said, extending their hands to Billy, Phyllis, then to me.

"Really pleased to meet you," Sean said, bowing as one of his balloon breasts fell out.

Several others attempted to step on it and, after a few tries, it burst. He pulled out the other balloon, stomped on it, and turned to Phyllis before bowing again.

This caused the remaining boys in line to approach Phyllis and Billy, bowing with one hand across their stomachs and the other behind their backs. Meanwhile, the remaining girls attempted their best curtsies to honour them.

These gestures were genuine expressions of respect and acknowledgement. Lives in Sahtlam, or Tansor, or on Lake Cowichan Road were hardly glamorous. Here were two well-dressed, attractive women in beautiful clothes and sparkling jewelry. It was an enchanting moment, and we all felt enriched by its magic.

Before we left, Barbara joined us and introduced the other teachers to her stepmother and second cousins, Billy and Kathleen.

Phyllis said, "You better hurry off and change into your jacket and tie before your father comes. If Mr. Garnett still has issues with whatever went on, you'll have much more influence as head boy if you're dressed like one."

"Good idea. And thank you."

"Thank you, Jeremy. I've thoroughly and enjoyed meeting your fellow students."

"And you, Phyllis," said Billy. "Run along, Jeremy. On the double."

I hurried to the change room, and the jocks cheered as I went in. They boasted about how much skin they'd felt in the brief interlude when the lights went out. Adolescent boys, predominantly, if not exclusively, heterosexual.

"Pipe-wrench," who was the captain of the rugby team, said, "Let me help you with your zipper, Jeremy. Gee, I could almost kiss you. Brothers, have you seen such a lovely shoulder, such a turn of the calf?"

I met his gaze, and within a nanosecond, we knew I would repeatedly reflect on that glance.

"Sorry, Jeremy," he said, "no cigar."

One of his rugby buddies added, "Or, in this case, pipe-wrench."

We all thought this was funny. The remark carried a quality of innocence and an element of empowerment. We had collaborated across barriers of sport and theatre association, gender identity, and industrial-rural farming origins, and had a great time.

By the time I walked past the office, my father was talking to Kathleen and Mr. Garnett.

"Jeremy," Dad said, "don't forget the crystal punch bowl."

"Sure, Dad, it's in the home economics room."

"I'll meet you in the car, son."

"Goodnight, Mrs. Long. Sir..."

I nodded to Mr. Garnett, who was still waiting in the hallway, but thankfully, had no words to convey to my father just how horrified he was with the scope and extent of the sneaky behaviors I had engaged in while dressed as a 1920s flapper.

# CHAPTER 18

**Tommy Comes to Town**

Sometime in the summer of 1967, while I was away at Expo 67 in Montreal, Colin Cameron, our long-time Member of Parliament, died.

As a family, we always went to the Fall Fair. That September, we ran into Dorothy and offered our condolences. Her manner was gracious as she introduced Dad, Freddie, and me to Tommy Douglas, explaining that we were members of "the Gabriel family."

She told Tommy that my grandmother and aunts were "stalwarts of the Women's Institute, and Fred's aunt, Sara Katherine, worked with Eleanor Roosevelt during the New Deal."

"This is a pleasure indeed, Mr. Gabriel," Tommy said.

"Fred," my father said. I put out my hand, saying, "Jeremy," which prompted my brother to add, "Fred Gabriel the Third, Mr. Douglas."

Dorothy went on to explain that my father's uncle, William, a steam engineer, worked for the Dunsmuirs and was horrified by how the workers were treated.

"Particularly the Chinese. He and his wife, Sara Katherine, assisted in organizing some of the fledgling unions in the area."

"I'm sure that made them quite popular," Tommy replied.

"Not with the Dunsmuirs," Dorothy said.

She said that when thugs hired by the Dunsmuirs went into the workers' houses and smashed all their furniture, Sara Katherine and other women she knew went in with hammers and nails and cobbled together the broken pieces of furniture so that the families could live with a sense of dignity. The pieces were referred to as "miners' furniture."

"We still have a piece," I said. "It even has little wheels on it."

"I'd like to see it," Tommy responded.

"Well, you're always welcome to come to our house," Dad said.

Dorothy stated she "would be remiss, if it wouldn't embarrass you, Fred, to elaborate on what happened to your aunt and uncle."

Dad nodded. "Go ahead."

She described how my ancestors were blacklisted and forced to flee as far as Portland, Oregon, to escape the influence of Dunsmuirs' thugs.

"As Sara Katherine was the ward of Sir Thomas Gabriel, Lord Mayor of London, and later ambassador to the Ottoman Empire, there was some connection to the Roosevelt family. While Franklin was Assistant Secretary of the Navy, Fred's uncle got the contract for manufacturing the boilers for the Liberty Ships, and they became quite wealthy."[8]

She went on to explain that during the Depression, Sara Katherine collaborated with Eleanor Roosevelt on a labour-intensive New Deal initiative to grow and process flax in Oregon, ultimately negotiating a contract to make the table linens for the White House.

Tommy appeared fascinated. "Tell me more."

Dorothy told him that Sara Katherine had received a Certificate of Merit from the Unemployed Citizens' League for valuable service rendered in the interests of humanity and aiding the destitute and unemployed.[9]

"And that's not all. My friend, Phil Thomas, the labour ethno-musicologist, believes that Woody Guthrie, the folk singer, sang 'The Ladies Auxiliary' to her when she received the citation."

"Very impressive indeed!" Tommy said.

Dad offered to arrange for Tommy to come to a Pulp and Paper Workers of Canada local meeting.

"You should also meet up with Sir Philip," Dad added. "Sir Philip?"

Dorothy elaborated for him.

"Sir Philip Livingstone was general surgeon to George V, but His Royal Highness, Prince Philip, as I've been told, didn't get along with him, so he and Lady Livingstone came to live here in the valley. Sir Philip is very progressive, a strong supporter of the National Health Service, and a proponent of the Fair Share Movement. We'll have to arrange for you to meet him. But Lady Livingstone, putting it politely, is somewhat arch.

I remember when we visited their home, and she showed me around the ballroom. On the railing encircling the second floor, there were paddles attached. She said they were the oars from a rowing team in an Oxford-Cambridge boat race "manned by men who became captains rather than captives of industry."

Shortly thereafter, Tommy was at our house with a campaign photographer. My father and younger sister were photographed in the downstairs carport of our new house, which was under construction.

I soon became a Young New Democrat and spent as much time as I could away from school in the campaign office and out in neighbourhoods, canvassing for support. My high school was one of the first in the country to have its own closed-circuit TV station.

Since I took TV production and spent a significant amount of time on the campaign, I negotiated credit for Social Studies 11 in exchange for a half-hour interview with Tommy on a subject relevant to Canadian society.

Tommy and I collaborated on an interview exploring American corporate ownership of the Canadian economy.

That conversation enabled me to do an infomercial for the Tuberculosis and Christmas Seal Society on the hazards of smoking. The promotion won an award, and my teacher, Jevington Tothill, and I attended an award banquet at the Hotel Vancouver in the city.

When I returned to the campaign office on a Friday afternoon, I learned that arrangements had been made for Tommy to have dinner with the Livingstones. Mr. Tothill and I had come directly from the ferry, and because he was a liberal (later, briefly, the head of the BC Liberal Party) and the best teacher I ever had, he dropped me and my bag off at the campaign office.

An immediate crisis arose when it was discovered that no one knew how to get to the Livingstones' home.

"I know how to get there," I said.

Someone called my mother and asked if I might accompany Tommy and his campaign manager to dinner. Dad was working the afternoon shift, and Mom worried about how I would get home. And did I have a clean white shirt?

The deputy campaign manager, Ed Whelan, a big man from Saskatchewan, gregarious and charming, got on the phone and assured her that a contingent from Crofton would be at the dinner. With my mother's permission, he would arrange for me to travel home with JH and Billy, his new friends.

"Fine with me," my mother said, "but stress to him to use his manners, and remind him, irrespective of having just won an award in Vancouver, and we *are* proud of him, not to be the centre of attention. He's only there to be supportive in getting Tommy to dinner."

Ed gave me a thumbs-up. I hurried to the washroom and changed into my pale green Edwardian-cut suit, a coordinating pale green Oxford-cloth button-down shirt, and what I thought was a smart tie.

When I came out, the campaign staff applauded.

"Hurry! Get in the car."

"Wait a minute," Ed said. He took a new tie from his desk

drawer. "Try on this one. The width of the tie should match the width of your lapels, and the bottom tip should touch the centre of your belt buckle."

He watched as I undid the collar buttons, tied the tie, and said, "Perfect! You're worth your weight in gold, son."

The driver would drop us off and then go to Victoria to pick up Dorothy's grandson, Robin. We had to ensure we were in the right place before he left. Tommy, Mr. Whelan, and I drove out in a Studebaker, with me sitting in front giving directions to the driver.

Tommy passed an oxblood-coloured leather valise to me.

"Hang onto this, Jeremy. It will give you something to do with your hands. If anyone asks you anything, just nod, look over at me, and say, 'Tommy probably has an idea or two about that.'

"There will be some important people at this event, and Ed has told Sir Philip that a Young New Democrat will be accompanying us. Sir Philip assured us there's enough food to go around. You can have a place with the kitchen crew in the breakfast dining room, so don't be anxious about anything."

I felt confident that the evening would go well and was honoured to accompany Tommy and Mr. Whelan. When we arrived, Sir Philip was already waiting at the top of the driveway. Soon after, other cars began to arrive, and he showed them where to park on the lawns.

After us, the first to arrive were JH and Billy with Mrs. Whittaker-Ramirez and Eileen Kathleen O'Malley, all dripping in style. While JH and Billy shook hands with Tommy and Sir Philip, Mrs. Whittaker-Ramirez approached, held me by the shoulders, and kissed me on both cheeks.

"My, my, you look like a bobby-dazzler."

Eileen Kathleen groaned. "Oh, Grandmother, don't flip your wig-zipper, it's only Jeremy!"

"Eileen Kathleen!" Mrs. Whittaker-Ramirez said reproachfully. "Please remember our discussion on the use of slang in polite society."

To set a formal tone, I extended my hand to Eileen Kathleen.

"How do you do, Eileen Kathleen, and may I introduce you to my friend, Mr. Whelan, the deputy campaign manager."

"How do you do, Eileen Kathleen, and this is your grandmother?

She nodded and smiled.

"My science and music tutor, Mrs. Whittaker-Ramirez," I added. "She is also an inventor and a linguist."

"I am honoured to meet you, Mrs. Whittaker-Ramirez." Tommy moved from Billy and JH and shook Mrs.

Whittaker-Ramirez's hand.

"And you've brought your lovely daughter."

"Granddaughter." Eileen Kathleen said..

The next car arriving carried Barbara Thornton-Sharp, her

stepmother, Phyllis, Kathleen and Eric Long. I knew Phyllis liked a certain precedence in introductions, and I introduced her first to Tommy. Phyllis, in her high German dialect of English, told Tommy she was "honoured and humbled" to make his acquaintance.

Meanwhile, Eric and Kathleen explained they were very much looking forward to supporting Tommy's campaign in any way they could.

Dorothy Cameron arrived with a tall, grey-haired man who immediately began shaking hands and introducing himself to everyone.

"Hello, Guy Binch, Brentwood College."

"And who do we have here?" he almost sang when he saw me. I extended my right hand and smiled.

"Hi, I'm Jeremy, Young New Democrat with the campaign."

"Hello, Binch, I'm Chopped Liver!" Eileen said, offering her hand.

Everyone laughed. Mrs. Whittaker-Ramirez looked at me with a hint of despair or resignation and rolled her eyes.

Dorothy swiftly rescued us.

"Jeremy, you're looking so dapper," she said. "I understand you've won an award for interviewing Tommy on television. We're so indebted to you!"

"Hi, Mrs. Cameron. Always good to see you."

"Dorothy, Jeremy. We're both working on the campaign.

First names for everyone."

Eileen Kathleen made her way to Sir Philip. She put out her hand and curtseyed as I explained to Dorothy that the award was for an anti-smoking commercial for the Tuberculosis and Christmas Seal Society.

Sir Philip told Eileen Kathleen that the curtsey, "nonetheless flattering is not necessary. We don't stand much on the ceremonial in these parts."

Mrs. Whittaker-Ramirez nodded to him, and he took her arm.

"Shall we go in?"

I was the last to enter the house. Guy sidled up to me and asked if I was the actor who had played Sir Humphrey in the Beaumont and Fletcher play, *The Knight of the Burning Pestle*.

"I am," I replied.

"I saw the performance at the MacPherson," he said, "and I was very impressed with your work on a particularly difficult part. You work with John Getgood and Pat Boulanger?"

"I do."

"I wish you were at Brentwood."

Eileen Kathleen rescued me from Guy's flattery offensive. "I'm sure you do, Binch!"

"Mr. Binch to you, my dear."

"Oh, we don't stand much on the ceremonial here," she said, mimicking Sir Philip.

"My bright young lady, do *not* confuse ceremonial cultural expectations with the clear expectations of acknowledging someone with the appropriate honorific, Mister."

"And I would remind *you* that Dorothy, one of the most esteemed Socialists here, stressed that Jeremy call her by her Christian name. Let's compromise." She smiled at him. "You may call me Eileen Kathleen, if I may call you Guy. Deal?"

"Deal, young lady."

Lady Livingston and her lady's maid swept into the foyer. While everyone else greeted her, Barbara, JH, and Billy came over to congratulate me on the award. Billy said I was "attired in a costume certainly more *au courant* than the charming 1920s costume I wore the last time we met at an event of some consequence. But wasn't that fun? Hopefully, tonight will be as much fun."

The next thing I knew, the maid was efficiently hustling Eileen Kathleen and me into the kitchen area, which was a hive of activity as three cooks and two male servers prepared the dinner. She knew both our names, called us Master Gabriel and Miss O'Malley, respectively, and asked if she may take my valise. I hesitated, as I knew it contained Tommy's notes and likely important papers.

"You're Barbara's nephew, are you not?" she asked. "Why yes, I have an Aunt Barbara."

"I'm a friend of your aunt and a friend of the Filbergs. You may be assured the valise will be secure and, should you need it, just ask George."

She called one of the servers. "George, please ensure Master Gabriel's valise is in a visible and accessible place in the breakfast room."

George led Eileen Kathleen and me into a sunny yellow room with stunning windows that immediately caught my attention. They made an impressive wall of diamond-shaped, leaded glass.

"You may wait here until you're called in with the kitchen crew to hear the grace being said. I also understand that if Mr. Douglas chooses to make a presentation after dessert has been served, you may be invited to sit with the staff along the wall in the dining room. Very democratic, wouldn't you say?"

After George left, Eileen Kathleen and I spoke to each other like old friends, the conversation flowing easily, as if no time had passed. She told me that since moving to Victoria, she'd been on several dates and had had four serious boyfriends, "each one better than the last," she said with a knowing smile.

One of them had gone to Brentwood College. She'd also heard a rumor that Guy Binch was enamored of the scholarship boys. She didn't elaborate, and I didn't ask.

"If he offers you a theatre scholarship, tell him you're an anarchist. No, you're up to your eyeballs with the socialists so he wouldn't believe that. Tell him you're a Marxist; that will scare him off. That is, if you want him to be scared off."

I didn't know what to say, as I was shocked that Eileen Kathleen would malign Mr. Binch. I saw him as a fellow actor.

"Eileen Kathleen," I said quietly, "I'm who I am because I've known men you might call all kinds of names, queers, pansies, fairies. Even Oberon, the King of the Fairies."

"And from where I'm sitting right now, I can tell you, those names are hurtful."

"The queer men I knew were honourable. They were my friends. And I might not have survived the bullying and the bullshit of high school without them. They showed me that there were still gentlemen in the world, whether they were artists, bankers, or carpenters. Men with dignity."

"They were wholly innocent and undeserving of slights of any kind. And I won't be part of any persecution or prejudice."

"I hope you won't be either." Her hand reached across to me.

"I get it. I'm sorry, and I'll be nice to Guy. I saw you in *The Knight of the Burning Pestle*. You were very good. Sir Humphrey was a fag, right?"

"Sir Humphrey was affected."

"You played him well, Jeremy. My boyfriend at the time said you were the highlight of the show. But I must comment on your suit.

Edwardian style? Impressive. I love the fabric, the texture of the light weave, and the colour corresponding with the shirt and tie. As I said, I'm impressed! Now let's get out of here."

Not knowing what Eileen Kathleen was going to suggest, necking, petting, smoking pot out in the yard, I had no idea and was more than a little anxious.

"I've eaten here before," Eileen Kathleen said. "My grandmother and Sir Philip are friends and talk about complex things. The last time it was about the need to establish a statistical discipline called radiological epidemiology to monitor radiation poisoning. The food is always good here; it's an art form. Let's go in and watch the cooks. They're probably about ready to present the food."

Drawing on her warehouse of charm, Eileen Kathleen and I went into the kitchen.

She began with, "I won't be so indulgent as to ask whether we may be of any assistance, but may we observe as you plate the serving dishes?"

The head cook smiled.

"It would oblige me to describe what is on the menu. The tardy guests have just arrived, and the others are finishing their cocktails. But the hors d'oeuvres are about to be taken in, and I'll ask you to follow George and Mordecai."

"Ladies," she said to her two helpers, "shall we go in with the plates of hors d'oeuvres? That will be the signal for someone to say the grace. We'll stand along the wall. I'm so glad you came in when you did, Miss O'Malley. Sir Philip said you were to be invited in for the grace, and, Lord, love a duck, I may well have forgotten, although that wouldn't have ruined the evening. Likely, Sir Philip and Lady Livingstone would have thought you'd said grace among yourselves."

Eileen Kathleen took my hand so that she would be first behind the waiters and said to me in a stage whisper, "Not bloody likely, comrade."

When we entered the dining room, the guests milled around the tables, and we noticed four more guests, Hiram Talbot, Harold and Vivienne Brown, and Lil Postgate. Vivienne wore her pearls but had her hair done in an elegant bouffant. She gave a brief wave.

Hiram Talbot opened his hands and mouthed, "Congratulations. I'm so proud of you!" Then he touched his heart."

I loved old Hiram, but I felt embarrassed about drawing attention.

"Welcome, Eileen Kathleen and Jeremy!" Sir Philip said. He confirmed with the servers that everyone had assembled.

"Not that we are going to ask Tommy to sing for his supper," he said, "but before, or while, the menfolk adjourn to sample fine brandy, we would welcome a rendition of Robbie Burns."

"It would be my pleasure," Tommy replied.

"And as you are the ranking theologian here, Tommy, can we prevail upon you to share with us the Social Gospel grace adopted by the CCF?"

"Initially adopted by the CCF," Tommy replied, "and occasionally revised, for inclusiveness, by various conferences and sections of the New Democratic Party. Shall we prepare ourselves in silence for the acknowledgment of this meal and those we share it with?"

We all remained silent as Tommy led us through the grace.

"We are thankful for this food here before us because of the collective efforts of those who grew it, harvested it, transported it and prepared it. We realize these treasures are our common heritage and what we wish for ourselves, we want as well for all others, so therefore, Creator, may we take our part in the world's works, joys and struggles."

After the chorus of amens, the kitchen staff, the servers, Eileen Kathleen, and I turned as if we were in a drill-team manoeuvre and marched into the kitchen.

The cook gave instructions to the servers.

"Give them four minutes, not five, with the hors d'oeuvres. Some guests were late. Then bring in the salmon, vegetables, browned potatoes, and the saffron rice dishes. Master Jeremy and Miss Eileen Kathleen, you two will have the first grasp of any hors d'oeuvres if any are left."

Before Eileen Kathleen and I returned to the breakfast room, Cook described the salmon, which was huge; except for the head and tail fin, it was peeled and the head had daubs of black olive tapenade for eyes with cartoonish, mustard-yellow pickle pupils.

The fish was served on the biggest silver tray I'd ever seen, resting on rosette swirls of poached Swiss chard, leaves and all. Around the salmon, there were wavy slabs of stuffing with pieces of dill, asparagus, shrimp, and red peppers. Scattered randomly over and around these were multi-colored cherry tomatoes and wedge slices of lemon. Two silver jugs sat beside the tray.

Cook said they contained "the sauce for the fish, two, because one was made with a base of smoked oyster oil blended with poached spinach. The other is a white sauce, as someone at the table may be of the Jewish faith and wouldn't want the oyster-flavoured sauce."

"Mordecai," she called to the second waiter, "will you stand behind Mr. Talbot and ask if he would prefer a white rather than an oyster sauce?"

Four minutes passed before George and Mordecai brought in the salmon. They returned promptly to bring in the remaining chafing dishes, working in quiet coordination. Only after the main course items had been arranged along the sideboard did they retrieve the trays of hors d'oeuvres from the dining room and bring them back to the kitchen.

About ten pieces remained, mostly delicate tartelettes filled with a lightly sweet curried fish mixture of crushed pineapple, coconut, and salmon. There were also a few crackers topped with what Cook called, rather proudly, "my sweet olive tapenade."

The two other cooks came to sit with Eileen Kathleen and me. They served us a second, smaller salmon with chard and tasty stuffing, which Cook called a 'terrine,' and asked if we wanted oyster or white sauce.

Both Eileen Kathleen and I said we'd try both. When our plates were served, Cook squirted the tasty sauces over our salmon and stuffing in a zigzag manner from plastic containers.

To make the terrine, one blended fresh chopped dill with two kinds of parsley, finely chopped celery, a quantity of chopped green onions, asparagus tips, whole grain breadcrumbs, a quantity of softened cream cheese, olive oil, sprinkles of Parmesan cheese, and as much poached shrimp and crab as was available.

The mixture was placed in a roll of cheesecloth, set inside the filleted fish from "stem to stern," and the fish wrapped in aluminum foil, shiny side in. The stuffed fish was ready to be placed in a hot oven.

She continued in a teacherly tone, precise and matter-of-fact:

"Taking out and taking off. No need to oil the foil. Removing it right after the salmon comes out of a hot oven helps peel the skin clean off the fish."

"The main body, with the cheesecloth-covered stuffing still inside, is lifted out using spatulas and set flat on a cookie sheet. The stuffing goes straight into the deep freeze."

"Meanwhile, the rest of the salmon is arranged on beds of hot Swiss chard rosettes, then tented with the same foil to keep it warm. Once the stuffing has cooled, we bring it out and cut it into rounds using a wavy cutter. Very handy tool, that. The rounds are then placed around the salmon, alongside the terrine, garnished with cherry tomatoes and lemon wedges."

"If we weren't expecting someone Jewish at dinner, I'd take the squirter and zigzag sauce over the terrine. But in a great house, one known as much for hospitality as for fine cuisine, it's a mark of a skilled cook to accommodate everyone."

"You don't see many Jewish people in these parts, but in England, among the gentry, it's not uncommon. Some have Jewish friends, business partners. Some had family who suffered terribly, or were killed, in the last war. I'll always go out of my way to make sure they're not served shellfish, if they choose to avoid it."

By the time she'd finished, we'd cleaned our plates and assumed everyone in the dining room had too. We were wrong. Eileen Kathleen and I were allowed to follow the cook to the dumb waiter's window, and we arranged ourselves on stools to watch and listen to what was going on.

Sir Philip and guests on one side of the table could see us, but after Sir Philip gave Cook a thumbs-up signal, she, Eileen Kathleen, and I were able to follow the conversations.

From what we heard and what Eileen Kathleen told me, Tommy emphasized the essential need for the separation of church and state in any free and democratic society. It was a contradiction that, in the mother of all parliamentary democracies, Westminster, the church remained Established.

"And the greatest and most recent victim of this was Her Royal Highness, Princess Margaret."

This sparked a discussion on divorce, leading to a consensus that legislation should be enacted to make divorce easier.

Someone asserted, "There should even be classes for anyone contemplating marriage, focusing on the rights and responsibilities of equal parties before a wedding is contracted."

Lady Livingstone commented that Princess Margaret had recently visited Africa and was reputed to have come up with a bon mot or two about royalty.

"Her Royal Highness was visiting an African village, and before the luncheon, an aged Chief showed her a photo of him with her Uncle David seated on an elaborate rattan chair, the Chief standing behind him. Margaret suggested they find the old throne and re-enact the photo, this time with the Chief sitting and her standing beside him."

"During the luncheon, various community members searched the abandoned thatched houses (*rondavals*), and ultimately located the ancient chair. When they brought it to where the official tour photographer was setting up her equipment, it was noticed that the chair was worm-eaten and likely unsafe for the old Chief to sit on."

"Princess Margaret was disappointed, as she felt the picture would be an appropriate gesture toward recognizing 'the role of monarchy in a post-colonial era.'"

"She was told it simply wasn't safe and that it would be bad PR if the old Chief were injured. To which Her Royal Highness responded, 'It just goes to show, people who live in grass houses should not stowe thrones.'"

Everyone found this witty, and Cook, who had listened with us, went off to recount the story to the rest of the kitchen crew and supervise the dessert preparation.

"The remark was not originally HRH Princess Margaret's," Phyllis said. "In fact, Her Most Gracious Majesty Queen Mary had expressed it originally when lawyers for the aviatrix, Beryl Markam, sued the Windsors for an annuity after it was proven Markham had

had affairs with David, the uncle of Her Majesty, Queen Elizabeth, and his brother, the Duke of Gloucester."

Tommy responded, "What? Levirate among the Windsors?" This got everyone laughing.

Sir Philip, growing jollier, said, "Be that as it may, and although this may seem heresy to some among us, I believe the current Queen was divinely ordained. I am aware that her uncle, David, had a condition shared by other descendants of Queen Victoria, specifically, without surgery, their testicles would not descend. In addition to hemophilia, Alexei, the son of Nicholas and Alexandra, suffered the same condition."

He stated with certainty that David had the same condition, which, while not causing impotence, rendered him infertile.

"And then there was the American Nazi minx who was a hermaphrodite," JH said.

"Technically, yes," Sir Philip replied, "but enlightened medical opinion is now considering her a transgender male-to-female person."[10]

"But she was also a Nazi collaborator?" JH asked. "Wasn't she intimately involved with von Ribbentrop?"

Tommy replied, "There was concern she was intimately involved with Joachim von Ribbentrop,[11] and the Privy Council knew that he and Molotov were involved with the Hitler-Stalin non-aggression treaty, so of course, concern and attendant caution were abundant."

I turned to Eileen Kathleen, who was all ears.

"Do you remember the last time Jake-Jack-John got sent to a foster home?"

"Huh?"

"Yeah. He told the same story about the Hawthorne boys. Said he read it in a confidential file on Sir Somebody or Other. He must have perused one, or likely more, of Sir Philip's files."

"Oh, probably. But shush. Listen to what they're talking about now."

Tommy was answering Vivienne Brown's question about the party's policy on euthanasia.

"I clearly understand the issue. It is truly inhumane for those with incurable illness to suffer the pain and anxiety we wouldn't subject our animals to. However, rational discussion on any related policy or practice is currently unthinkable due to the influence of Rome on the Liberal Party."

Sir Philip argued the point made.

"Aren't we supposed to have a division of church and state? In a modern liberal democracy, physicians and their patients should have this as an officially recognized medical treatment option. And it would likely receive Royal Assent. Again, bringing up Her Most Gracious Majesty, the late, great, Queen Mary, it is generally known she consented to euthanasia so as not to impose an official period of mourning that would interfere with the scheduled date of the Coronation."

Phyllis rose suddenly. "I must excuse myself."

Eileen Kathleen said, "Jesus, Mary, and Joseph, this is better than Peyton Place!"

"This would be an appropriate time for the men to retire to the study," Lady Livingstone tactfully suggested. "Ladies, shall we have our dessert in the salon? Let's catch up with Phyllis."

May my granddaughter join us?" Mrs. Whittaker-Ramirez asked. "She has been practicing a Mozart piece."

"By all means," Lady Livingstone said.

"What about Jeremy?" Dorothy asked. "Can he join us? I'm dying to hear about his experience with television production."

"By all means."

"Sneak in with the men and we can compare notes afterwards," Eileen Kathleen whispered to me.

Her suggestion was made moot by the lady's maid who came in.

"Ah, young people. You have been invited into the salon to have dessert with the ladies and to hear a musical presentation by Miss O'Malley."

On the way through the dining room, Hiram took my arm.

"Your grandmother would be most proud of you for being supportive of Tommy," he said. "And I'm proud of you, too, Jeremy."

In contrast to the dining room's sociability, the communication in the salon seemed awkward to such an extent that the contrast between the two settings was striking. Phyllis was noticeably shaken by the disclosure that Queen Mary had consented to her own death, but she quickly rallied.

"I had previously heard it as Court gossip, but I hadn't registered the realities of it until this evening."

Ignoring the comment, Lady Livingstone said she was "very impressed with Mr. Douglas, and pleased that, as a Baptist, he understood prohibitions against levirate."

"After all, we are members of the Anglican Communion because Henry VIII had to form the Protestant Church. The Pope of the time erred in allowing Henry's marriage to Catherine of Aragon, as she had previously been married to his older, frail brother."

"Sir Philip and I will be supporting Mr. Douglas because of our own self-interest. In a democratic society, there must be parliamentary interests to ensure hope for fairness and reform.

Otherwise, we could easily find ourselves in a situation where the poor would revolt against us and we would lose everything, not least our lives."

"Let's hear some Mozart!" Dorothy suggested.

Eileen Kathleen was happy to perform, followed by her grandmother, who played a Chopin nocturne.

"I should like to hear the Moonlight Sonata," Phyllis said. "I've already played it," Mrs. Whittaker-Ramirez replied.

Phyllis shrugged. "It must be what the French call, déjà écouté," adding, "I own that I'm in something of a state of shock over the information concerning our late Sovereign Lady, Her Most Gracious Majesty, Queen Mary."

"Phyllis," Mrs. Whittaker-Ramirez said, interrupting, "getting back to your suggestion, I play the sonata, it was likely a case of morphic resonance, triggered by recall of your time with Queen Mary."

Dorothy rescued us from this drollery. "No spoon-bending, please! Jeremy, tell us something uplifting about your latest theatrical success."

I welcomed an opportunity to hold forth.

"After the awards banquet, I linked up with some of the other winners, a few of whom were involved in musical theatre in Vancouver, and we all knew the Sister Suffragette song from Mary Poppins."

Lady Livingstone looked interested. "Suffragists, they had a song?"

"Several!" Dorothy replied enthusiastically.

"May I consult with Mrs. Whittaker-Ramirez?" I asked. She, Eileen, and I found the key, and the next thing we knew, we'd lightened up the room.

From Kensington to Billingsgate One hears the restless cry From every corner of the land Womenkind arise! Political equality and equal rights with men.

We paused, and Eileen Kathleen placed her hand over her heart.

Take heart, for Mrs. Pankhurst has been clapped in irons again.

"A scarf! A scarf!" I shouted.

Dorothy, quick on the uptake, tossed me a large doily, which I draped over my head with exaggerated flair.

"No more the meek and mild subservient we, We're fighting for our rights militantly. So don't fear…"

Eileen Kathleen and I launched into an animated march around the salon, arms pumping, feet in step.

"We'll cast off the shackles of yesterday, Shoulder to shoulder into the fray, Our daughters' daughters will adore us, And we'll sing in grateful chorus, Well done, Sister Suffragette!"[12]

The room burst into applause, even Phyllis clapped, her composure momentarily undone. Eileen Kathleen and I gave theatrical bows.

"Jolly good!" Dorothy said. "My only criticism is the term 'Suffragette.' I think it trivializes what was a serious and brutal struggle. Can we sing it again, and we'll all try to join in but, this time, let's say, 'Sister Suffragist.'"

We started again, and when we were halfway through, a knock at the door revealed JH and Guy Binch standing with a flashlight. JH said that Tommy's driver was en route from Victoria, bringing Dorothy's grandson. A telephone call had been received "asking for someone to stand at the end of the driveway with a torch so they wouldn't miss the turn."

Tommy suggested Jeremy do this, and Guy said there may be cougars in the area. He would accompany Jeremy, but I thought three of us against a cougar, particularly if we were singing, would scare off any critter.

"Fine with me," I said. "Ladies, it has been a delight."

I went around the room, acknowledging everyone and nodding to Lil Postgate, who said, "I hope to see both television productions."

Lastly, I approached the piano and kissed Mrs. Whittaker-Ramirez on both cheeks. Eileen Kathleen curtseyed, gaining a laugh from the women.

"Remember what I told you, Jeremy. Goodnight, Mr. Binch, lovely to meet you. 'Night JH."

As we passed the kitchen, I remembered the folio. "George, the folio!"

He pulled it from a shelf and bowed.

"Please extend my appreciation to Cook and all the staff who made Eileen Kathleen and me feel so welcome."

He bowed and nodded again.

"It was our pleasure, Master Jeremy."

As JH, Guy, and I hurried down the driveway with the flashlight, JH started singing.

When I was a lad, I served a term "As an office boy to an attorney firm."

At the end of the song, Guy said, "I was tasked this evening by the headmaster to invite Tommy to speak at an assembly. He has agreed to speak at Brentwood, and I've invited him to bring a group of Young New Democrats to watch our dress rehearsal of Antigone. He agreed *if* the school would promise to deliver a busload of students to a hootenanny with the Young New Democrats."

"What, pray tell, is a hootenanny?" JH asked,

"It's the latest thing, some kind of folk-song jamboree."

"What is Antigone?" I asked.

"A play," Guy replied. An Athenian tragedy by Sophocles about the consequences of challenging the power of the state. Antigone and her brothers were the children of Oedipus.

A car approached, interrupting us and flashing its lights. JH shone the flashlight, and the driver, a trade union activist from the Lower Mainland, emerged from the car.

"Sir Philip," he said, "Gerry Stoney, IWA."

"I'm flattered, Mr. Stoney," Guy replied, "but, regretfully, I'm not Sir Philip. I'm Guy Binch from Brentwood College."

He turned to JH and put out his hand. "Sir Philip . . ."

"Strike two, Brother Stoney, I'm a mere ideologue and theoretician, but you'll be pleased to know I always stand with the workers."

"Hi Gerry," I said, "glad you found us."

Dorothy's grandson, Robin, the other passenger, got out of the car.

He was tall, blond, and dressed in an unassuming yet wholly presentable manner. His hair, longer than any man's I had ever seen, gave him a rock-star quality. He carried a presence that I imagined his grandmother had in her youth: healthy, strikingly attractive, and possessing remarkable eyes. It felt as though their knowing gaze contained a thousand perspectives.

Was I being judged by this privileged prince of the Party? What must it have been like to grow up with a grandparent who was a Member of Parliament? And to have someone as worldly and wise, resourceful, purposeful, and as confident and witty as Dorothy for a grandmother? Was he sizing me up? He had likely encountered enough other well-dressed, ambitious Party activists to determine who might be helpful or who might have their own agendas.

I would rely on conventional manners. Upon learning of a death, I knew it was important to approach the bereaved swiftly, to offer consolation, read their emotions, and remain present until they had acknowledged the spirit of the gesture. I extended my hand.

"Welcome, Robin. I'm Jeremy from the Young New Democrats. I've known your grandmother for years. Let me extend my belated condolences on the death of your grandfather. My family knew him, supported him, and always spoke highly of him."

"Thank you, Jeremy," he replied.

To no one's surprise, Guy was next to greet Robin with an enthusiastic handshake and a query.

"Are you to be part of the hootenanny cast or chorus?"

Robin glanced at me questioningly, and JH resolved the confusion.

"Guy was just explaining that Tommy has invited some of his students to a hootenanny in Duncan in support of the campaign."

Gerry added, "My union will be supporting a couple of top-notch folk singers from Burnaby who will be coming next month."

"Well, that clears things up," Robin said. "For a second, I thought I was being asked if I belonged to some obscure caste."

We laughed, and Guy replied, "Tommy has also committed a group of Young New Democrats to attend our school's production of Antigone."

As we walked up the driveway, Robin remarked, "Tommy doesn't usually speak on behalf of the YND."

"Are you familiar with Antigone?"

Robin nodded, amusement flickering in his eyes.

"Not personally. One of Oedipus and Jocasta's children was King Priam's granddaughter, actually, Priam's great-granddaughter, as her mother was simultaneously Antigone's grandmother and her mother."

He recounted this with such ease that it sounded as though he was describing one of the children of families who lived on a neighbouring Saskatchewan homestead.

"Sophocles, wasn't it?"

"It certainly was," Guy answered.

Lowering his voice, JH told Gerry, "Oedipus was Antigone's half-brother as well as her father."

Gerry looked genuinely concerned that introducing decent YND members to this sort of theatre may have political consequences.

He put his hand on my shoulder and asked, "Are you okay with this talk?"

"Oh, sure," I replied. "I know all about Oedipus. I grew up in Crofton. Half the tomcats in town are called Oedipus."

Everyone laughed as I took the light and led the group up the driveway. Halfway there, Robin stopped.

"Two-legged or four-legged?"

"Only the four-legged," I assured him. "Cats and dogs."

"I'm pleased to hear that."

"Me too," Gerry said.

We arrived at the door, and George announced, "Gentlemen, into the lounge, there is a glass of port awaiting you."

"Porteau, George, porteau," Guy said, "A Portuguese fortified wine."

Gerry explained his refusal, saying, "Sorry, Brother, I'm driving."

Robin expressed his wish to see his grandmother. "I'll get her," I said.

Assuming his entry to the salon would be like introducing a

charismatic force into the measured grace of altar guild, I tapped on the door and opened it a crack.

"Dorothy, you're needed!"

She sprang up, anticipating Robin, and I felt slightly embarrassed as they hugged one another. "My darling!" Dorothy exclaimed.

My darling grandmother," Robin murmured.

George directed me to the study and, as I entered, Tommy greeted me wholeheartedly.

"The man of the hour!"

I turned to see if someone else had followed me in. "Matters of consequence, Jeremy," Sir Philip said. "Guy and Tommy have conspired to radicalize the sons of the haute bourgeoisie, and you are to play a part!"

Hiram interjected.

"With respect, Sir Philip, they are likely the sons of the comprador bourgeoisie. The haute bourgeois send their boys to Shawnigan Lake."

"Now, Gerry," Tommy said, "will you profile this hootenanny event for Jeremy?"

I was directed to one of the vacant leather armchairs, and we listened with interest as Gerry explained how a church hall could be transformed.

"Candles in wine bottles, folk singers with guitars, the usual kind of event. After a few Bob Dylan, Joan Baez, and Donovan songs, Tommy will speak briefly. Just a few words, once the mood is right."

Then, once the crowd's warmed up, Phil and Hilda, two first- rate singers from the Vancouver Folk Song Society, will take the stage. They'll do some labour songs, starting with 'Where the Fraser River Flows.' "Should stir the place up nicely."

He might have gone on, but Sir Philip genially interrupted. "The stuff of a planning session for sure."

Ed Whelan took the cue and thanked Sir Philip for his gracious hospitality, and Tommy went with him to say a final good night to the women.

I sat with JH and the other men as they discussed how far we'd come as a movement and how we would greatly benefit from having Tommy as our MP.

Dorothy, Robin, and Lil Postgate came in and were offered refreshments. We were all invited back to the dining room, where extra chairs had been brought out. Cook, George, and Mordecai served desserts, a cream trifle filled with various fruits and brandy- soaked pound cake, clay ramekins of crème brulée, and coffee.

Eileen Kathleen attempted to charm Robin. A twinge of jealously crept in, and I nearly leaned in to whisper, "She's had four boyfriends since she left Crofton," but held my tongue.

There was competition around who would drive me home, but I left part of myself in the breakfast room, the salon, and the study lounge. Like Cinderella, returning from the ball, I felt as if I had met a prince, or perhaps two or three, and was brimming with stories to share with my parents and anyone willing to listen.

# Chapter 19

**Ideological and Other Challenges at the School of Social Work**

One consequence of my unpaid undergraduate practicum at an international development assistance consortium was unexpected. When the ideological crisis occurred, I remained insulated from its impact, simply because I was the only unpaid person.

I ended up there because I had been in India on a United Church of Canada youth exchange. As well as touring numerous development-assistance projects, I spent time in Kashmir by day, riding horseback on scrawny ponies in the foothills of the Himalayas. During the glorious nights, I stayed on a houseboat on Dal Lake.

In Calcutta, home was the ecclesiastical palace on Free School Street. All this occurred during the Emergency when Prime Minister Indira Ghandi announced the Twenty Point Programme, which included the statement: "Economic offences are the worst offences against the poor and will be punished most harshly."

In the summer, immediately prior to my senior practicum, as a CUPE member and social-work student, the flexible schedule of my work at the university's convention centre allowed me to be a student delegate to the UN-Habitat Sub-Conference on Social Welfare and Human Settlements.

I met official delegates, accompanied them to informative proceedings, attended the Habitat Forum at the old Jericho army base and, literally, sat at the feet of Margaret Mead, Mother Theresa, and Buckminster Fuller. They were there to describe the environmental, social, cultural, and economic impacts of the plutonium economy.

When the long-simmering ideological conflict finally erupted, the board of directors took action, ultimately leading to the dismissal my clinical supervisor and everyone in the political animation collective. However, I couldn't be fired, as I wasn't officially an employee. I was a student and had a contract.

It was generally known by most board members, particularly the social democrats, that I was enamoured with and worshiped my clinical supervisor. Yet, everyone involved acknowledged that my supervisor's pedagogical investment in me was entirely platonic. When board members maligned him as "Mr. Radical Chic" or dismissed his Marxist-Leninist analysis as mere Mao Tse-Tung thought, I remained silent.

Instead, I took on the assignments that had previously been handled by my supervisor and other dismissed members of the animation collective handled. This was necessary, as the agency had received several hundred thousand dollars for deliverables, which included extensive travel for development education presentations to colleges, faith-based organizations, and labour unions.

Well-dressed board members met with professors from the School of Social Work, negotiating time away from my classes, essentially enabling me to save the agency. Everyone recognized that serving as a participant-observer in change agentry and systemic transformation would be an invaluable learning experience. I was, however, required to attend my weekly Directed Studies Dialectics seminar and to complete major assignments for all my classes, as well as fulfill my contractual practicum responsibilities.

This meant that one of my fellow students would earn academic credit by collaborating with me to outline the coursework expectations for each of my classes. This wasn't an onerous expectation, as we took the same classes.

It also offered my friend and fellow student an awareness of the dynamics and developments of ideological conflicts and related negotiations. A confidentiality agreement was negotiated to ensure that nothing I told my confidante would be discussed with anyone at the school.

The problem that emerged due to our top-secret liaison involved an assignment to be completed between the fall and spring semesters. Susan, my friend and student colleague profiled the assignment as "conceptualizing a letter-writing campaign on an issue of social import."

I work full time and then some at the international development consortium, travelling extensively by plane, and taking on weekend shift work as a resident attendant at the university residence. Life was an organized blur. I had the confidence of a board member who had been appointed acting supervisor. This allowed me access to agency meetings, except those involving terminated employees, and relevant documentation about anything else impacting the agency.

Late on a Thursday evening after I'd flown in from a meeting in Terrace with Stan Persky, a community college adjunct professor I had known through the Company of Young Canadians, I sat at my desk. Meanwhile, board members filed out to recharge their energies in preparation for making it through Friday.

I viewed a stack of files, several marked "Priority," and recalled Stan's advice: "Try to synergize. Do purposeful work that benefits both the consortium and the School of Social Work. Remember, you're uniquely positioned to raise the academic community's awareness of development, politics, and the oppression of that privileged intellectual class sanctioned and funded by the State to engage in agitprop."

One of the priority files was from the London Committee for Human Rights in Latin America. Mindful of Stan's advice and aware of my list of academic assignments, I took a breath, surveyed my office, and thought, *I'm so glad I'm here.*

The file contained an urgent action appeal directed at the Canadian Minister of Manpower and Immigration. The request sought approval for the Minister to vacate an Expulsion Order concerning a Chilean family of ten.

The family was at risk of detention by military and paramilitary forces supportive of the Pinochet junta. It also highlighted the need to secure status, provide material and other forms of support, and settle them within a supportive community.

I worked on the appeal throughout the night. By morning I had edited and re-edited a two-page profile of the family. It detailed their involvement in community and social development projects promoted by the Popular Unity Government of Salvador Allende, as well as the reasons each parent and all their children were at risk of detention, torture, and disappearance. I also wrote a covering letter to the prime minister, the minister of manpower and immigration, and, as a courtesy, my MP, who happened to be the minister of finance.

I wished the prime minister and his family the best of the season and forthcoming the new year. I also respectfully requested he turn his attention to the circumstances of another family at risk of living in vulnerable conditions in a country where democracy, as we enjoy it in Canada, does not exist.

Similarly, I conveyed my compliments of the season to the ministers and their families and requested they review the attached profile of a vulnerable family. For humanitarian reasons, I further requested that each of them bring up this family's situation and the proposed remedies with the prime minister during their appointments with him before returning to their ridings for the festive season.

The acting supervisor, Diana, was the first into the office. "You didn't stay here all night!" she said.

"I did. It was one of the priority files. Can you read it?"

"What say we read it over breakfast, then you can go home and get some sleep. I understand you work afternoon shift tonight."

"Graveyard shift. I don't need to be at the university until midnight."

She moved toward the door.

"I'll be back in a minute," she said.

The time was considerably longer and enough time for me to make six copies of the letters, one for the prime minister and each for the ministers, one for the agency file, one for my social-work assignment, and one for the London Committee for Human Rights.

Breakfast was at a tiny café beneath the Fir Street exit of the Granville Street Bridge. I ate while Diana read, and I watched her, trying to read her facial cues.

The proprietor came around and asked, "More coffee?"

"Yes!" she said. "Would you please leave the pot, and don't disturb us for at least half an hour?"

With a brimming cup, she launched into her review.

"This time we're spending together constitutes official supervision. Before we go, I want to hear about where you were, Prince Rupert? Terrace?"

"Both," I replied. "I also renewed the contract with Stan Persky."

She picked up the file of letters.

"This is brilliant," she said, leaning back with the paper in her hand. "It shames us all and our entire enterprise that this was almost overlooked."

"Did you know that someone from our MP's office has been asking about 'rumours of turmoil' and whether we're 'doing any international refugee advocacy work?' I have a meeting with her today."

"I'll take this and ask that she fax them with a covering note saying, 'Originals posted in Vancouver on this date.' Before I go to the meeting, I'll recopy these on heavier bond, and we'll make a copy for the media if we need to take further action."

She set the papers down carefully, already mentally ticking through the steps ahead.

"May I come?" I asked. "No."

"I could answer any clarifying questions."

"No need. You've done an excellent and timely piece of work. I'll let this Liberal busybody know that as well as doing legitimate advocacy, we're also supporting the work of a talented and principled social work student. I'll ensure the originals are mailed today; they don't need stamps."

"I know that."

"What else don't you know? Come on, I'll drive you home. Get some sleep, and again, thank you for all you do. Everything else is going okay."

"A full day off? Everything *is* wonderful!"

***

I worked the afternoon shift at the university residence on December 23rd. Chuck, a fire fighter from Kamloops was staying at my apartment, as his niece was in Children's Hospital.

I'd left him with my private line at the university in case there was an emergency, and when he called, I was talking with international students who had no place to go for Christmas.

"I picked up your mail," he said, 'and you've got a telegram. Shall I drive it to where you're working?"

"No. Just open it."

"What if it's personal, you know, confidential?"

"Yeah, you're probably right."

He said he'd see me in half an hour. I also told him I'd call in an order to the Chinese food place in the UBC Village and asked if he could stop and pick it up.

I was playing Christmas carols on an old baby grand piano in the Place Vanier student lounge when Chuck came in. He lingered by the doorway for a moment, watching as the African students and I finished the Huron Carol.

When the last notes faded, I turned to introduce him. "This is Chuck," I told everyone, "the kind soul who delivered the telegram."

I could feel the curious eyes in the room as I broke the seal and hurriedly unfolded the paper.

From Bud Cullen, Minister of Manpower and Immigration, it read:

The Expulsion Order has been vacated. STOP All ten members of the family are safely in Canada with full refugee status. STOP. They will be supported to locate in a Canadian community.

STOP Congratulations on your efforts to effect this outcome are acknowledged and appreciated. STOP Merry Christmas to you and loved ones. STOP "Wow!" I shouted.

There were outbursts, handshakes, and cheers. Agnes Moletsane from Lesotho asked if we might offer thanks. She led us in a prayer for the family, thanked God for my efforts, the school of social work, and for my roommate in his role as messenger.

I called Susan, and her mother, a president of an Ontario Chapter of Hadassah, answered the phone. Susan wasn't home, but she'd be pleased to take a message. When I told her I'd received a telegram from the Bud Cullen, she interrupted me.

"I know Bud. You want me to tell Susan word for word what he wrote?"

"If you don't mind."

"I don't mind. Go ahead." I read the telegram.

"Mazel Tov! Good work, Jeremy. I'll leave a note where Susan will see it when she get's home. Anything else?"

"May I wish you the best of the season?"

"Is it too much for you to say Happy Hannukah?"

"Not at all!"

I went further and gamely sang.

Give me oil in my lamp keep me burning, give me oil in my lamp I pray; give me oil in my lamp keep me burning, burning, burning keep me burning 'til the break of day.'

Happy Hannukah, Mrs. Gutnick!"

"Where in the world did you learn that?"

"India. On a United Church of Canada Youth Exchange."

"Well, that is delightful indeed. And Happy Hannukah to you too," Jeremiah.

Susan called before we finished the Chinese food, and she sounded pleased. The students, Chuck and I sang more Christmas carols until it was time to return home.

Information about the letter-writing initiative wasn't received well by my professor, who assigned and graded, or, more accurately, refused to grade the effort. Susan told everyone in our Directed Studies Dialectics seminar, and the professor, Ben Chud, asked, "Jeremy, how are you?"

I explained that the social-advocacy professor had said that I needed to do another assignment, and that my efforts were unworthy of a mark since I had misinterpreted the assignment's instructions. I was to *conceptualize* a letter-writing campaign and wasn't to operationalize any form of initiative. In doing so, I had risked causing impacts to the school, the university, and my own enrolment.

"And he went on to say, 'Central to understanding social work practice is understanding agency.'"

He paused, lips pressed tightly together, as if weighing every word. Then he spoke.

"Agency," he said at last, his tone clipped. "And with no

sanctioned agency from the school or the university, you have written to the highest office in the land."

The words hit me like a physical blow. I wanted to protest that the result of my efforts had been socially productive, but his expression told me it didn't matter.

"You have one week," he continued, "to resubmit a revised approach to an epistolary advocacy initiative. Your grade will depend on the quality of that new work." He glanced at his notes before adding, "It will be marked with a ten percent demerit for being late, and in consideration of the number of gross errors in judgment and probity represented by your actions."

As I recounted this reprimand to my dialectics class, a flash of heat spread across my face. Humiliation pressed down on me, joined quickly by anger, and then by a cold ripple of fear for the consequences. My eyes burned. I almost cried, which would have been worse than anything, here in front of my beloved professor and my closest classmates.

"Did he say anything else?" Ben asked.

"Just that, as a student, I'm entitled to make mistakes, a component of the learning process, and that in time I may reflect on this experience. I should be thankful I had learned, 'without harm to any human population,' the value of understanding the critical role of agency sanction in social work endeavours."

"Before I invite discussion and analysis," Ben replied, "I must clarify that it isn't my usual practice to comment on the professional practice of a colleague, but there are exceptions to the rule. I also want to bring to your attention, Jeremy, that at least one of the things said by my honourable colleagues is clearly wrong. Anyone?"

Susan spoke first.

"He consulted with his clinical supervisor, and his placement resides with the sanction and support of a community agency, I might add, where he's not being paid."

"Relevant points," Ben said agreeably, "but the specific relevant point is this: Jeremy didn't write to 'the highest office in the land,' unless I'm missing something. Jeremy. Did you copy the Governor General?"

"No."

"Well, we have cause for an appeal," he began, folding his hands on the table. His tone was measured, but there was a glint of determination in his eyes.

"Information with which you were admonished and denied a mark. whatever the reason, it was inaccurate." He leaned forward slightly. "You appropriately appealed to the most senior decision makers who have statutory sanction to approve your request."

He paused, letting that sink in before continuing.

"And, as a courtesy, you informed your MP, who, in his role as finance minister, would have a direct or indirect interest in being informed of the potential economic costs of the effort."

His gaze moved slowly around the room, inviting engagement. "Anyone else have a perspective?"

Peter, whose mother was also a social work professor said, "I would like to stress my wholehearted support of Jeremy's appeal, should he decide to take that course of action. Is there not some value, given factors that may pose a risk, such as the fascist government of Chile that compels us to take action in situations like this?"

"That is called a defence of necessity," Ben said, "and the principle is two-fold. First, policy is there to guide the wise and restrain the foolish; second, it's often easier to ask forgiveness that to obtain permission."

He looked around the room.

"Let me stress that I'm not suggesting forgiveness is even remotely relevant to the matter."

I appealed the mark. Peter and Susan came with me. The professor wrote that he had responded without the knowledge of my having consulted with my acting agency supervisor.

It was his "sincere hope" that, when engaging *actors* who occupy political office I think twice, and twice again, before documenting any matter that might currently or in the future be subject to a freedom-of-information request.

He further acknowledged, because of my location in an international development agency, that I had knowledge of the situation in *Chili.*

I neither corrected him nor his assertion that I'd written to the highest office in the land.

I secured an A+ on the assignment.

# CHAPTER 20

**Another Assignment**

Looking back, I can see how my part in the ideological crisis opened doors I never expected. Graduation was barely behind me when a consortium agency offered me a position, and with it, a move to Montreal.

This position took me to Montreal, where I began taking courses at McGill in *Jeux Linguistique*, a series of strategies designed to rapidly acquire the dialectical nuances of any language.

Over four months, both at McGill and in other learning environments, I became sufficiently fluent in the Selangor dialect of Bahasa Malaysia to participate in meetings within the Malaysian Federal Ministry of Culture, Youth, and Sport. I was even authorized to speak with the media on official matters.

Upon returning to Canada, I was hired to oversee a community development project at the Burnaby-based Community Centered College for the Retired.

Officially, my staff and I were paid to create a handbook of community services for Burnaby's retired population. Unofficially and in numerous ways, we brought in resource people, distinguished scholars, and an interdisciplinary team to an elementary school facing declining enrolment.

Our aim was to challenge ageism by reconfiguring the social

constructs affecting the retired population within a specific urban North American jurisdiction.

Unofficially, and with the full approval of Lin Latham, the formidable president of the college, I also worked with a group of her friends and fellow travellers from an organization called The Voice of Women.

Together, we worked to mobilize faith and community groups in opposition to the development of an American nuclear submarine base in Bangor, Washington.

Shortly before World War II, Lin was enrolled in a master's degree social- work program at Harvard and engaged in a practicum with the American Society of Friends. Although an Anglican, she embraced the Quaker commitment to non-violence.

After Pearl Harbour was bombed, any Quaker involvement would risk her status as a foreign student. She left for India, while there, she fell in love with and married 'Jolly' David, an officer in the Royal Navy.

During her coordination of an outdoor concert for the Bombay Symphony Orchestra at the India Gate, she arranged signals, by semaphore, for Royal Navy ships to fire their cannons eight seconds before the cannon segment in Tchaikovsky's 1812 overture.

When Lin and her friend, Dr. Pauline Jewett, then president of Simon Fraser University, learned I had visited the Ghandi Institute for Nonviolent Studies, they encouraged me to join the efforts to train community members in organized nonviolent civil disobedience.

This work involved liaising with an organization called the Pacific Life Community (PLC) to train community-based affinity groups. Their efforts proved to be quite successful.

I lived at the time in the old Winch house at Nelson Avenue and Imperial Street in Burnaby. The Winches were prominent in the CCF and had long before vacated the three-storey home. It even featured a hidden staircase, historically used to discreetly sneak people in and out of meetings.

Through arrangements Lin made with Simon Fraser University, I had the opportunity to live there alongside a group of feminist academics who were members of the Black Uterus Collective.

After publishing the *Burnaby Seniors' Handbook* and completing the contract with federal funders, I turned my attention full-time to organizing opposition to the proposed submarine base.

Guided by insights from Margaret Mead, a diverse coalition came together, members of the Pacific Life Community (PLC), the Voice of Women (VOW), Greenpeace, and the environmental agency SPEC joined forces with the American Indian Movement (AIM) and several Canadian unions.

Together, we began a campaign to expose the connections between uranium mining, nuclear power, and the escalating arms race.

I was collecting Unemployment Insurance and, after a year, my benefits ran out.

By this time, I had moved out of the Winch house, and David,

Taeko, Greg, and I had settled in Earth Embassy, a communal house on 8th Avenue near MacDonald in Kitsilano. It served as a clearing house for coordinating community groups involved in the Bates Royal Commission of Inquiry into Uranium Mining.

We also obtained money from the Student Association of Simon Fraser University to unite church groups, rural and urban community organizations, union activists, and others in hosting a multi-faceted Nuclear Awareness Week at SFU.

I was also serving on a global planning committee for an international day of anti-nuclear protest. We received honoraria for speaking engagements and funding from several organizations, including the Women's International League for Peace and Freedom and the Environmental Health Committee of the BC Medical Association.

Even so, Earth Embassy's finances were stretched so thin that finding spare change for postage stamps was a challenge.

I began applying for social-work jobs with the provincial government, submitting scores of applications but receiving no responses. When I told Lin, she simply said, "Leave this to me."

Lin spoke to Pauline. She and I had attended a World Federalists Conference together and she liked me. At that conference, Pauline reacted to the right-wing, corporatist, but simultaneously laissez-faire proposals about international development presented by retired television personality, Stanley Burke.

"I've always thought that anyone who asserts such trivia must either be ignorant, insane, or evil. I must stress though, Stanley, I've never considered you insane."

She suggested to Lin that I write to the Provincial Secretary, Grace McCarthy, and inform her that my applications were not receiving any responses. Lin said I could use her name as a reference and include a copy of the Seniors' Handbook. "And keep me informed. If nothing happens Pauline might pull some strings; she really respects you."

I wrote to Provincial Secretary McCarthy, and it worked. I received a call recommending that I apply for a specific job competition in Lillooet. In researching the community, I learned that a social worker friend and colleague had also applied for the job.

We met and strategized, read all the relevant Acts, obtained all the demographic statistics on the town, and talked to community members. Everyone wished both of us the best of luck.

Finally, I was invited to an interview. A Salish friend who had just returned from Ottawa lent me a white shirt. I pressed my dress pants, chose an understated, tasteful tie that complemented a brown-grey-green Harris tweed jacket, and arrived on time.

Along with two child-protection supervisors in the room, there was a personnel technician and a high-level public-service member. In the middle of a question by one of the child-protection supervisors, the public-service official waved his hand.

"What in God's creation does working in inter-racial youth integration in Malaya have to do with working with Indians in Lillooet?"

"Is this a query to determine how I manage conflict?" I asked.

"You stop right there. Does your hubris have no bounds?" I turn to the supervisors.

"I supervised the field placements of professionals and non-professionals in all manner of community-based agencies in four specific states."

"I told you to stop right there. What you're selling we aren't buying. Do you think we have no idea on what you're up to?"

"Enlighten me, please. What am I up to?"

"This interview is over. Do you have anything to say for yourself before you leave and note, anything you say will be recorded."

I paused, took a deep breath, and blew on my hands. When I felt sufficiently poised, I said, "Just so you don't embarrass yourself further, the country you refer to hasn't been Malaya since its independence in August of 1957, and this sister member of the Commonwealth of Nations is known as Malaysia."

"Note in the documentation I have had to repeat, this interview is over."

*** 

A week or so later, I learned my friend had got the job. She already had rented a house in the community and graciously invited me, saying I was welcome to stay as long as I wanted.

She also promised that each night, on returning when from work, she would clarify the names and functions of various office workers, the organization of files, the expectations of file recordings, and the names of forms.

By that time, I was in shock, and I thanked her and said I'd think about it. For any number of reasons, I wanted a conventional social-work job. I wanted to start making some money, join a union.

I felt increasingly burned out working as an international anti-nuclear activist. I loved what I was doing, but the accompanying poverty was eroding my health as well as my judgement.

Two weeks later, a letter arrived from BC Public Service informing me that no one suitable had made application for the referenced position, and the competition for social work employment in Lillooet was being withdrawn.

It was a punch in the gut. I went to my room and wept. This was unlike any grief and loss I'd ever known. After investing years of volunteer effort, study, tuition, and enjoying a taste of success and influence, I felt totally disempowered. I could do nothing but weep.

I experienced a *reactive* depression. Intellectually, I knew my intense grief was logical; emotionally, I felt powerless, even when I tried to strategize. Should I call Lin? If I did, she'd tell Pauline, who may or may not call Grace, the provincial secretary, and suggest some conflict-resolution process, stressing there had been a human-rights violation.

If that occurred, resulting in a job, it would be a compromised way of getting into the Civil Service. It would likely impact assignment and promotion, and I could end up in remote areas such as Atlin or Lower Post.

Intellectually, I resigned myself to never working in the BC Public Service. Emotionally, I was devastated. I loved my work and knew it was important, but I wasn't making any money.

Taeko arranged for me to meet a Japanese graduate student struggling to write her thesis on the Dependency Theory of Superstructural Economic and Social Development. I met Fusako Honda at her West End apartment.

I agreed to work with her two afternoons a week for three hours to transliterate her thesis into a suitable submission for her thesis committee. And she paid me well.

In addition to the many hours we spent reviewing text and footnotes, after each afternoon session, Fusako took me to one of several Japanese restaurants where we revisited the day's work and plan for our next sessions.

I encouraged her to spend time each week participating in a cultural project that she could document in English, then read to me over our dinners. In this way, she refined her ability to think critically and discuss subjects in conversational English.

She already had a master's degree in library science, but initially, she could not provide or describe context. That changed with the jeux linguistics and language enrichment activities I was able to engage her in.

I accompanied her to an art exhibit and had her describe

individual paintings and drawings in the Fenwick Landsdowne Birds of the West Coast collection.

Taeko also arranged for a couple of friends to give me a makeover, a manicure, pedicure, and facial. The boyfriend of one of my feminist friends, Tang See Hang, took me shopping to buy the right clothes so that I might unobtrusively appear at Fusako's swank West-End apartment or accompany her to exclusive restaurants.

One afternoon, while I prepared to leave for Fusako's, David came into my room.

"Is your passport still valid?" he asked. "Why?"

"You need to go to an organizing conference in Columbia, Missouri, then Washington, D.C., then maybe elsewhere. Maybe Manila for that Westinghouse reactor project we've been monitoring."

"A group of lawyers associated with Ernesto Prudente from the Polytech Institute has a report they need to get out of the country.

They can't trust the mail and don't know anyone else who can courier it to the International Philippine Patriots' Association. Are you up to it?"

"When?"

"The day after tomorrow. Sisters of Silkwood want you there. VOW, the Environmental Committee of the BC Medical Association, SPEC, and a lawyer who wishes to remain anonymous will pick up the bill. But you need to get two volumes of the Biological Effects of Ionizing Radiation report from the Library of

Congress."

"That's all?" I asked

"There's a bit more involved."

That afternoon, I told Fusako I would be gone for a while. She gave me three chapters to take with me, and I left her with culturally enriching assignments to enhance her critical thinking and refine her English grammar and pronunciation.

I had arranged for her to attend a monitored language lab at UBC. Prior to leaving that afternoon, she gave me a cheque for $200, much more than she owed me. I agreed to fax completed edits of her chapters.

In person and remotely, I worked with her for another two months. After she completed her thesis defence, she went back to Japan, and I later learned she had taken a job with the United Nations, first in Jakarta, then in New York.

I wouldn't have survived the rejection by the public service if it weren't for having had such an ardent and intelligent student. I couldn't have presented myself as a graduate-student tutor were it not for Taeko's friends giving me manicures and facials, and Tang, knowing my size, bringing me preppy shirts, slacks, a linen sports jacket, and various pairs of shoes.

How did I meet Leila and Tang? Several years before, I had organized tenants in Richmond during a transit strike. One morning, I picked up Leila, Tang's girlfriend. She was hitch-hiking to her job as the office manager of a posh local spa.

I still drove my Sunbeam Alpine at the time, and she put her large bag behind the bucket seats. En route to Richmond, we shared small talk, and when I let her out on a busy corner, she grabbed her oversized bag.

It wasn't until I got into the car later that afternoon that I realized she had left her purse behind the bucket seats. And quite a purse it was, made from an aardvark.

When I opened it, I noticed her office keys, her wallet, and a small card that read, "If ever this is lost or found, please call me." I called the number, and a man answered. I explained that I had found a purse in the back of my car and asked how I could return it.

"Don't hang up," he said. "Leila! The man in the sports car just found your purse."

The next voice was hers.

"Hello, this is Leila! I can't believe this. It's a miracle! We couldn't open the spa this morning, and everyone said I'd never see the money or the keys. A locksmith is coming here this evening and to the spa. Thank you!"

"Can you tell me where you're located?"

"Oh, yes. We live just off 12th Avenue near Burrard." She gave me the address.

"I live nearby," I said, "but I'm still in Richmond. I can be there in about half an hour. Don't worry. Nobody broke into the car. Everything in the purse is safe."

She hugged me when I got there, and I could see she had been crying. She had called the spa owner to let him know the purse was being returned, and he instructed her to take me to a restaurant of my choice and reward me in any way I asked.

"Oh, none of that is necessary," I replied. "Being an honourable person is its own reward."

"Yes, it is necessary," Tang said. "In my line of work, international finance, we don't meet people like you. Leila tells me you are a community organizer who works with childcare centres and tenants' unions in public housing. And you returned a purse with a float of at least one thousand dollars in it? Everyone we talked to today told us we'd never see the money. Where would you like to go to eat?"

Neither of them had ever been to the Normandy between 10th and 11th Avenues on Granville. My friend Annie McGeachy's mother, Dorothy Sherrard, ran the place and gave me a warm welcome. Over dinner, Tang offered that if ever, for "two days, two weeks, two years, or two decades, we can ever be supportive of you to repay you for this, just ask."

"The reward for me is meeting you two. And, of course, dinner. Thank you for being so gracious."

Annie's mother, Dorothy, came to our table. "So how do you know Jeremy?"

Leila explained, and Annie's mother, in her thick Queensland accent, said, "That's what I would expect of him. Jeremy is a friend of my son-in-law, and I would expect the same of him or any of my children. Did he tell you he is from two old Vancouver Island families from the Cowichan Valley? Cowichan Valley people, in my experience, are all generally like this."

After whatever happened with the public service rejection, I went to Leila and Tang's place, but they had moved. When I visited Dorothy at the Normandy to ask if she had seen them, she said, "They come by from time to time. They're doing very well."

I told her what had happened and burst into tears. "If you see them, could you ask them to contact me. Just tell them I'd like to visit with them, and I don't need anything but their company."

"And maybe a good meal. You sit right here until I get back to you. Today's special is roast beef and Yorkshire pudding. Now don't tell me you've gone vegetarian. Dinner is on the house."

# CHAPTER 21

**Sad News of Jake-Jack-John**

I had no intention of having a good time in Washington, D.C. But in spite of my prejudices, generously handsome radicals took me to charming bookstores, favourite places, DuPont Circle, impressive buildings...

By this point in my life, I had spent several years organizing demonstrations in cities across the country against corporations involved in nuclear power and nuclear weapons. During the protests against the Trident Nuclear Submarine Base and with the encouragement of my friend and clinical supervisor, Lin, I led a roving conflict-resolution team on the day before the first UN Special Session on Disarmament.

That day, our work drew more attention than we could have imagined. The next morning, we were on the front page of both The New York Times and The Wall Street Journal.

Because we were able to profile the Trident submarine system as having a first-strike capability, delegates at the Special Session decided to conduct a second round of deliberations with a broader scope. After this success, I found myself with an itinerary that included travel to places I thought I would never visit.

It was the peak of the arms race. The world had reached a "balance of terror" that exceeded the "balance of power" with the achievement of global overkill capacity. At home, alongside others, particularly my housemates at Earth Embassy, I had been organizing efforts to halt uranium mining in BC.

I also found myself serving on a Global Planning Committee, coordinating an International Day of Protest during a conference in Washington.

While staying in a sumptuous apartment in Chevy Chase, I called home and learned from my father that they had heard news of Jake-Jack-John.

"What has he been up to?"

I interpreted the ensuing silence as ominous. "What has he done now?"

"It's not what he's done," my father said, "it's what's been done to him."

"Which is?"

"I saw his father in Safeway, and he looked like hell, so I asked, 'How's the family?' And he told me Jake-Jack-John had been murdered. Executed, he said. It had something to do with a drug deal gone bad, and Jake-Jack-John is alleged to have informed to save himself. He was supposed to be part of a witness-protection program, but something went awry. Are you okay with hearing this over the phone?"

"I'm shocked," I replied. "But I'm not surprised. I always knew something like this would happen to him. I'm sad . . ."

Dad interrupted. "There's not going to be a funeral, so don't hurry home. Do you need any money or anything?"

"No. I'm fine," I lied, my whole body shaking.

"We'll talk when you get home, although his dad was pretty closed-mouthed about it all."

"Okay."

"Your mom is making a big batch of rhubarb cake, and she'll take some up to Winona at the hospital."

"What's wrong with Winona?"

"Nothing's ever wrong with her. She's got a job at the Nanaimo hospital doing occupational therapy, as an aide or trainee or some such role. Apparently, she didn't take any time off work."

"I have to go."

"You're going to be okay? I know you were close."

"Not for years."

"Well, old Jake-Jack-John was special."

"Dad, I must go. My dime."

"Next time call collect. Your mom wants to say something."

My mother's voice came through, anxious and quick. "Are you okay?"

"About Jake-Jack-John?"

"No, where you are, who you're with. Are they good people?"

"Mom, I'm fine, just a little sad. I'm with excellent people associated with an organization called Catholic Worker."

"It's always a little strange when a family chooses not to have a funeral. Come home as soon as you can."

"Is there something else that's wrong?"

"Only that you're in the most wicked, evil city on the planet."

"Mom, there are charming parts, and I'm meeting some really neat people."

"Just be careful and don't trust anyone there. Anyone."

"Mom, I'm okay."

"Be careful. I've got to get these loaves in the oven. Love you."

"Love you too, Mom. Love to Dad. 'Bye."

I left Washington shortly after that. After spending some time in New York City, I stayed in a pottery co-op in Vermont. Feeling depressed, I rationalized that I needed to return to Canada, attend meetings, and organize the liaison with Japan and Australia for the Day of Protest.

By 1981, the country was in a recession. I couldn't officially work anywhere while traveling to various locations to organize civil disobedience events. For a couple of years, I supported myself by coordinating childcare programs for New Democrat conventions and other meetings for progressive groups. Childcare consumes one's life. How I managed to get up to the mischief I organized, I still can't figure out.

One of the best programs I coordinated was during the fiftieth anniversary of the founding of the CCF. I cared for the grandchildren of individuals who had been on the site fifty years earlier to establish the party that introduced socialized medicine to Canada.

We had floppy-haired clown sprinklers and ice rock cairns, kids running National Film Board cartoons on the Canadian Labour Congress projector, and some wild parachute games. The children of politicians and Party activists I worked with, who had participated in previous convention childcare programs, were part of a helpful group called the UTLF, the Under Twelve Liberation Front.

I wrote a report at the time entitled "The Pink and Blue Report: A Working Document on Organizing Childcare for Large Political Gatherings" (1983). One section read:

I will not, for reasons of national security, describe the two- line struggle over a tent for older kids to have art play and meeting space separate from spaces normally occupied by infants and toddlers. I will say, however, I appreciated the way convention coordinators kept me informed on negotiations with park officials on whether we could even erect a tent. Finding a tent was also quite a problem.

After exhausting all Party contacts, the RCMP, the Canadian Forces through official channels while respecting appropriate protocols, and Saskatchewan Native Organizations, I refused to contact the Conservative Caucus. Instead, I called the Emergency Measures Organization and, using the daily code, arranged for a parachute to be flown in from Canadian Forces Base Moose Jaw by helicopter at no charge to the Party.

If, over the Canada Day weekend, we were to experience excessive heat, the kids would require some form of sunshade. I met the chopper at the Regina airport, as it seemed unlikely to have been well received by anyone if it wiped out one of the trees, or, God forbid, the bandstand or fountain in Confederation Park.

A thank-you note was sent on white bond, no letterhead. A copy was given to an informed Moose Jaw delegate who returned the parachute by car to the back door of the officer and gentleman responsible for the favour. Thank God for gay men in the military, at least when they're engaged in Search and Rescue or socially relevant EMO tasks, they're not practicing how to kill people.

The Socialist kids enthusiastically participated in Canada Day celebrations. When they visited the Museum of Natural History, their arrival corresponded with the July 1st Twenty-One Gun salute. I heard several versions of what happened from the three staff members who attended, as well as from some of the children who were old enough to describe their experiences.

The museum has a large lawn. Shortly after the staff and children exited their cars, the salute began. The group, startled, caught the attention of numerous onlookers by running amok across the expansive lawns. With each cannon report, they all fell dramatically, as if wounded by gunfire. Hurriedly, they helped one another to their feet and ran pell-mell until the next report sounded. Even the staff joined in the fun. All the children who attended this field trip received Canada Day pins and helium balloons.

The pins were used to pop the balloons, and there was enough helium for each child and most of the staff to carry on their conversations in the high-pitched, cartoonish tones of Daffy Duck, the F sound stretching into a comical "eeh-aaaf-FF!"

I also put together childcare events for Native organizations. The Cree and Chipewan peoples, concerned about uranium mining, had gathered at a place called Pine House Lake in Northern Saskatchewan. Since there were no other resources available that met my standards for quality childcare, we occupied the local health centre.

Local Indigenous peoples and health-centre staff treated me as though I were a social-development worker. After a multi-lingual planning meeting, which took the form of a long-term regional social needs assessment, a series of interesting workshops took place.

I had hoped that a rock musician I was smitten with at the time would appear so that we could have a commitment ceremony with traditional elders and wild young radicals blessing the relationship. That didn't happen, but the following did.

### Ke'Saskatchewan genderbender

I did childcare for a unity gathering

Of two historically feuding nations.

One-day workshop involved

Ancient midwives from several lands

Big young women and small old ones

275

Sat on chairs in a circle.

New Health Centre flipchart read

Birthing Workshop

As women talked

And translated long heroic walks

Through blizzards to cabins, hot summer

Births came slow, autumn births

Came fast, good times and fat babies

Bad times and sad women.

I lay on my back listening

Bouncing then-fat Aquat[13] on my legs from

Knees to feet til' she was sound asleep.

When I wasn't looking

A German woman took a picture

Of the flipchart and some old women

Aquat asleep, me with a beard.

When we put together the scrap book

People from several nations laughed.

Gentle teasings, "Squat for a while

And walk a-ways then squat again

Aquat, she'll come out headfirst.

We won't even make you shave

Or give you a hysterectomy."

I did childcare for organizations ranging from the Canadian Labour Congress Weekend Schools at Simon Fraser University to various faith-based groups. I even consulted on childcare for the General Assembly of the World Council of Churches and environmental groups.

After one such event, during which we effectively stopped the development of the Hat Creek Thermal Coal Generating facility, I considered undertaking an extensive retreat in a Benedictine monastery.

# CHAPTER 22

## Fasting at the Fish Camp

Kochie stared at me as if I'd just suggested dancing barefoot on a bed of hot coals. For an Interior Salish person who carried the weights of his people's history in every glance, my casual mention of working with anything tied, even faintly, to the Catholic Church was more than surprising. It was unthinkable.

"They'll get their hooks into you. You'll never be able to leave. They'll turn you into a zombie. You want to experience a residential school? You're still pretty, Miss Brazil, they'll see you comin', and they'll eat you up. Why not come and play anthropologist with me for a while? We'll feed you up good."

I made feeble attempts to explain the article of the Benedictine Code that forbids Benedictine orders from turning away anyone seeking refuge. None of the people I was with believed it.

I went to Lillooet for a couple of weeks, where my Salish friends introduced me as an anthropologist. For years, around the anniversaries of the bombing of Hiroshima and Nagasaki, I organized demonstrations, conferences, die-ins, letter-writing campaigns, and workshops.

Before I left for the fish camp, a radical nun I knew, Sister Rosalie Bertell, suggested that engaging in a silent fast while at the camp could be valuable.

I thought I'd give it a try. My hosts said I could write, pray, eat, but not speak, and I eventually wrote Salmon and sage and smoke and salt-brine Sun and stars and fire always burning; Sacks full of fish from the river below, Pauline Johnson whispers me images,

Swift as this summer night, dark as this river, Salmon and lavender, sun and smoke

By the river in silence, August the sixth.

Secure, into a silence fast I hear only

Rational stories of people whose face lines

Speak only the honesty and knowledge of history

Woven with suffering, warped with smoke

And strengths and salmon,

Sage in summer is sweet and strong.

I hunger for insight but see outside

Only a Fisheries Official in brown khaki

and spy glasses, perched on an old, very old

Officially condemned bridge. A significant thing

Enduring with purposed beside someone officially

Insignificant and merely glimpsing

Salmon and sage, sun and summer.

Lying low, watching him watch the Indians fish

I see how swiftly he surveys the sheds of salmon strips

Too fast to count, he gets no scents, no sense of style

Intricately sliced geometric designs

Solemn, elegant signatures he could never copy

Or identify.

From up at the camp, I watch him watch the Indians fish

And hear the deafening roar of August the seventh

Wincing at the invasion of boot prints in the dry

Stoney soil by the river, while the uniform, robot-like

Occupies the khaki man for whom the summer scene

Is only something measurable. He is too foreign to taste

Or enjoy, scents of wind and water in this sweet warm air.

Testing his footing on the perfectly sound but

officially condemned bridge, I wonder, what he's ever

Trusted, what he examines, who he obeys.

August, Caesar of months. How fitting that the state arrives in uniform, to spy on Indians between August the sixth and August the ninth. Like those broken days, for those broken Days and decent broken ways, Margaret Mead tore her book to pieces in three days, I break my fast and report to my hosts The specifics of the intrusion.

Ancient Eva asked me to sing 'I'll be seeing you in all the old familiar places' and roll your eyes like you mean it; I know what you're doing.

She tapped on my transistor radio, "the Japanese."

So I sing, and rolling my eyes, know. There's time. And time. Caesar's uniform time and the time for seizing the time, like this time, by the river, by the old bridge, the year it was condemned, by the pathways, Close by timeless generations, august and sage, intricately slicing Geometric designs, making solemn, elegant signatures in drying salmon.

One night, after I had started talking again, a couple of drunken rounders parked on the road above the camp and arrived with a couple of cases of beer. They wanted to buy fish. For several reasons, chief among them that the fish was being prepared for a strictly traditional fishery, I was elected to get rid of them.

"Don't talk to any of us," Koochie told them. "We don't want anything to do with you. We don't want your money or your alcohol. Talk to our anthropologist."

He turned to me and said, "Tell 'em off. That's your job."

I explained that I was researching how a traditional fishery was organized. They wouldn't go away, and they cracked open beer after beer while I stood my ground and refused to let them pass me. They said they'd go when they got either fish or women. When I said there were no women in the camp, they pointed at Eva, who was eighty years old. I warned them that if they didn't leave immediately, they would face legal consequences that would make their lawyers' heads spin for years.

For reasons unknown, they accompanied me to the road where they had parked their pickup. My friends watched from a distance as

I cajoled, intimidated, used sarcasm, and joked with them. They continued drinking and asked what a "fancy-assed person" like me was doing with "a bunch of Indians in the middle of nowhere."

I talked about Gandhi, how we were all kindred, and that it was necessary to see the connections between each other.

They looked disgusted. "Asshole," one of them said scathingly, "you don't have no connection with me. Your type doesn't know the shit people like us have pulled. We've been in so many jails."

Still claiming to be an anthropologist, I explained the ethnographic principle of six degrees of separation, that everyone on the planet has only six other people between them and everyone else.

"Fuck man, you don't know anybody I ever crossed with."

I kept on in the same unruffled tone. "I'm not wanting to intimidate you, but likely, I lived in a fraternity house with someone who represented you in court, or someone whose mother is a judge."

"I have never been up in front of any lady judge."

"Were you ever in Brannon Lake?"

"No lady judges there."

"Were you ever in Brannon Lake?"

"When I was a fuckin' kid."

"What about Okalla?"

"Okalla, BC Pen, Kent, Matsqui, Kingston. You know the warden or someone? Your daddy a guard?"

"Maybe his mother's a guard," the second man said.

"Nope. My mother's not a guard, but see, if we spent enough time together, I bet I could come up with someone we all know."

"I bet. Fancy-assed kid like you, you're not even a good archeologist. Your theory is full of shit. I think we just go down there and get whatever we want."

"Ever hear of anyone by the name of Jake-Jack-John?" I asked.

"Holy shit!"

"You knew that piece of shit? He ain't a friend of yours?"

"Six degrees of separation…"

"What?"

"The principle of six degrees of separation. I take it you're familiar with the case of someone known as Jake-Jack-John?"

"Fuck, crack me a beer. You knew that asshole? He was bad news. He was a piece of dog turd."

"You knew him?"

"Did you?"

"When I was a kid."

"Are you going to have a beer?" I smiled agreeably. "Is it cold?"

"Is it cold? Give Mr. Archeologist a beer here. Would you like a joint?"

"Oh, no. I couldn't possibly. It would endanger my

research."

"Do you mind if we imbibe?"

"Suit yourselves, but tell me what you know about the case of Jake-Jack-John."

"You wanna hear what happened to him? It turns my stomach to even think about it. You wanna hear?"

I nodded.

"You asked for it. Dudley, that was his name. Dudley gets out of Kent and gets into a halfway house. He has some community work in an organic garden with a bunch of do-gooders."

"He starts going to holy-roller meetings and gets paroled. What he was really doing wasn't robbing folks at the organic garden.

Instead, he was learning everything he could about hydroponics. He had friends from every prison he'd been in, and soon he was recruiting them to grow some of the most potent weed around."

"The shit is so good, it's stronger than the acid on the street, and sooner or later, someone gets busted and the whole growing operation gets raided, and the stuff gets traced back to the organic garden."

"He gets called in, and he doesn't wanna do any more time, so he sings. He names fucking everybody. Everybody connected to the shit going on, even some of the people at the organic garden, gets hauled in. Even a probation officer and her social-worker husband."

"So, he gets in witness protection again, but he figures he's the crown jewels and nobody can touch him, so he starts dealing from some other growing operation he has."

"By then, he was supposed to be spending weekends in a motor home with a gay cop who claimed he was helping him get a new identity. One long weekend in May, the cop was down on the Lower Mainland, and Dudley saw his chance. He took the motor home, drove up somewhere near Youbou, and packed it full of weed, ready to sell to anyone who showed up."

"Word gets out, and some of the boys out on parole go up there and catch him red-handed. He tries to talk his way out and says it's his mother's motor home, and the guys get all pissed off at him, pound the shit out of him, take his dope, take all his money, and piss off."

I listened, completely engrossed, then asked, "They left him there?"

"Let's say parties unknown came back."

"And?"

"He was still alive, and that wasn't good for anybody, least of all him, and there he was, or so they say, with his dyed blond hair and all. They figured they'd bleach his hair, but nobody had any bleach. So they got out a Yukon credit card and got the gas out of the motor home, then doused it all over him and the seats. Then he came around, and the boys were going to torch the motorhome with him in it but decided they'd have a little fun instead…"

"What does that mean?"

"You gotta understand, plenty of people were mad at him by

then, and it was almost daylight, or so they say. They cut his throat, and just to send a message, they grabbed some nails and a crowbar and drove a couple of spikes through him into the Arborite table.

Then they released the brakes and bailed out. They thought about tossing in a match, but when you've finished off a rat, no one sticks around. He was already gone. That was the end of Dudley."

The second man contributed to the end of the story.

"We don't know anybody who knew anything about any of this. It's just all talk. He's a goner, though. Nobody ever heard anything about him after that."

They traded theories like poker chips, each one wilder than the last.

"Whoever started this story probably made the whole thing up.

They wouldn't want people thinking the witness protection program isn't worth a damn. His family got paid off, no funeral, nothing."

"Maybe they gave him a new motor home, though."

"Someone's girlfriend, a nurse's aide, swore he survived, that they flew him out of the country and gave him a sex change. But no one's heard a word since, so I'm betting that was crap."

"Yeah? So he didn't get his throat slit? We'd better be going.

Good luck with your archaeology."

"Anthropology," I corrected, the word tasting strange in my mouth. My voice sounded far away, even to me. And for a moment, I couldn't shake the feeling that somewhere, somehow, Dudley might still be out there. Watching.

# CHAPTER 23

## Social Work in the House of Commons

I thought graduate school would be the biggest challenge of my year.
I was wrong.

Not long after classes began, my friend Joy Langan, fearless,
sharp-tongued, and a sitting Member of Parliament, called with an
unexpected offer. She wanted me in her Ottawa office as a policy
analyst. The task? Uncover the hidden regulatory history of silicone
gel breast implants. The goal? Nothing less than getting them
banned across Canada.

I had researched medically induced trauma impacting people
with AIDS. When Joy tracked me down and asked what I knew
about breast implants, I could have said, "Just what I've read in the
papers," but we went back a long way and had mutual friends.

"Joy, what do I know about breasts?"

"You'll be perfect! If I have a woman intern working on this, the
issue would be relegated to one of numerous issues impacting
women's health and equity. Still, with you, we'll identify the issues
on a much broader terrain."

I had hoped to get an internship at the National Research
Council, but the possibility of getting a top-flight security clearance
to work in the House of Commons was an appealing prospect. I
signed a contract and found myself with an office in Centre Block,

down the hall from the parliamentary restaurant.

There is nothing quite like working inside a legislature. It is a place where certain words, spoken in the right rooms, can shift the balance of power. I will not linger on the everyday intrigues of that world.

I never tried to pass myself off as a typical policy analyst, nor will I outline my exact standing in the Parliament Hill culture during the strange, waning days of the second Mulroney administration.

What mattered was that I looked the part.

And somewhere in those marbled corridors, the redoubtable dean of the Commons, the Honourable Stanley Knowles, took me under his wing. More than once, he leaned in and offered the same piece of advice: "Watch your back."

As I got to know Stanley, I felt it necessary to be circumspect in all my dealings with him. I often walked down the hallway at some godforsaken hour, carrying an armload of research, and Stanly would ask, "Want a cup of tea?"

He and Joy were both printers by trade. She had kept Stanley informed on the progress of her efforts to upgrade the moratorium on breast implants to an official ban. She hired me to manage the evidence dump in response to her Freedom of Information request to the department that initially had approved the implants for sale in Canada.

One afternoon, Stanley asked me if I realized the Medical Devices Branch of Health Canada was a department of the Radiation

Protection Branch.

I knew that was the case, but felt reluctant to approach the branch. My presence as one of the lead anti-nuclear trouble-makers was already well known. Still, for years, one of my life goals was to infiltrate the Radiation Protection Branch. I was determined to complete the internship, possibly even successfully finalize it, and officially ban silicone gel implants. Achieving that meant thinking at least twice before taking any action.

Stanley assured me I likely had a better structural analysis of the branch and its regulatory functions. He described its somewhat tarnished, or at least lacklustre (Stanley's term), relationship when it came to signalling concerns about the mandates, policies, and practices of the Atomic Energy Control Board (AECB), Atomic Energy of Canada Limited (AECL), and its export entity, Atomic Energy of Canada International (AECI).

"Schedule a meeting with the director and stress you're a student," he advised.

"Your faculty advisor will expect you to be thorough, get a complete departmental overview instead of cherry-picking the policies of a single sub-department. Start at the top and work your way down. Be unfailingly polite so no one can justify refusing to cooperate with a well-spoken representative from the office of as earnest and determined a politician as Joy."

"When you want to know something specific, say, 'The Member for whom I work would like to know . . .'"

"You'll probably get a bafflegab response, but take notes. If you can pull it off, ask if you might pose a further clarifying question. If too much is going on, call the next day to thank the director for the time he spent with you and for his generous hospitality." Then add, 'There are just a couple of things I'd like to revisit. May I ask at this time, or should I schedule a follow-up appointment?'

"He won't want to meet with you again, so what have you got to lose?" You want to find out which documents were excluded from your FOI request. You want to get that information.

Stanley would invite me into his office and ask all kinds of questions. Then, he shared stories about his childhood in Los Angeles or recounted how he had worked as a printer in Winnipeg. As a printer, he always knew what was happening with any progressive group that needed meeting notices produced.

I knew something about Stanley that he never mentioned, so I kept my counsel and waited for the right moment to raise the matter.

During one afternoon, I had the honour of accompanying him to a function in the courtyard of the east block of the Senate. The Member of Parliament for whom I worked was hosting a barbecue for the graduating class of the Labour College of Canada, and Stanley wanted to attend.

As he needed a Party flack to run interference for him, I had the role of standing with him, holding a smart, leather folio, and taking notes and contact information from people who wanted him to follow up on various matters.

I had a great time talking to people whose kids I had looked after while doing childcare for the Party. When it was time to leave, Stanley said, "Doesn't look like you got to eat much. How about dinner in the Parliamentary dining room?"

As we walked back to Centre Block, I said, I had to ask him something important, "and that it was a personal rather than a political matter."

"Isn't the personal political and the reverse also true?" I nodded. "Go ahead."

"Stanley, where I grew up, there was a woman who was a staunch member of the Party. She signed me up in the Youth Section when Tommy ran in the riding."

He looked interested. "Go on."

"When she died, she left my sister a strand of pearls and a set of pearl and diamond earrings. The case they're in is from a jewelry store in Winnipeg."

"Nice case?"

"Oh, very nice!"

"What colour?"

"Blue, velvet on the outside, satin on the inside. The woman I mean was from a small Icelandic community in Manitoba. She had married a man who returned from World War I with severe disabilities. To my mother and a few close friends, she confided that you had once wanted to marry her, but divorce was nearly impossible in those days. When she finally left Winnipeg, because she was in love with you, you gave her the necklace and the earrings."

"What was her name?"

"You mean you gave other women pearl necklaces and diamond earrings?"

"I don't recall anything as specific as personal relationships before 1981, when I had my stroke. What was her name?"

"Vivienne."

"I don't recall anything as specific as personal relationships before 1981."

By this time, we were in one of the elevators going to the dining room, and a huddle of senators came in. Senator Gigantes greeted Stanley warmly.

"Mr. Knowles, it's always wonderful to see you. Remember when we went to the Soviet Union in 1956? Those were the times, weren't they?"

"I'm afraid I don't recall much before 1981, when I had my stroke. But I got quite a chuckle over your very precise remarks on lackeys in the recent Senate filibuster."

A Conservative senator had accused Senator Gigantes of being a lackey. He responded by discussing a battle between Sparta and Athens, at which time Spartans were known, tribally, as Laconites.

When the ships of Athens invaded and filled the harbour, and the Spartans shielded themselves behind craggy rocks on the cliffs above, the Athenian commander shouted, "Laconites, you are outnumbered! Why not surrender to our superior force without any bloodshed?"

There was no response from the Spartans. The Athenian commander pressed on.

"If you do not surrender, we will darken the sky with our spears."

As Senator Gigantes spun out this slice of history during the filibuster, buying precious minutes, his voice carried a deliberate rhythm.

"And the Spartan response, the Laconite response, was the very essence of what we know of Sparta: sharp, hard, and uncompromisingly intelligent. Do you know how they answered? It was the pure distillation of the Laconite ethos, the Laconite identity."

He paused, letting the suspense build before delivering the punch line. "One of them simply shouted, 'Some shade!'"

"And the Honourable senator opposite thinks he is insulting me by calling me a lackey. I stand proud to carry on the brave, smart, strong heritage and ethos of the people of Sparta, the Laconites."

When we entered the dining room, Stanley was mobbed by the female servers. A few kissed him on the cheek, and a couple even kissed him on the mouth. I stood back, holding the leather folio.

When we were finally seated, a young French-Canadian waiter asked in a condescending tone, "Will there be anything special?" This was said with a particular pejorative emphasis on "special," which prompted me to ask, in my most arch-Cowichan accent, "How is your sushi?"

"We don't have any."

"Pity."

Stanley touched my forearm. "Well done."

The server then asked, "You want sushi well-done?"

"Touché!" Stanley said, and we all laughed.

I enjoyed our dinner immensely and, as we walked back to his office, he took my arm and said, "Vivienne?"

"Yes," I replied.

"Was she happy?"

"I think so."

"Good. Goodnight, Jeremy."

"Good night, Stanley, and thank you for a memorable evening."

***

Throughout the day, I conducted telephone interviews with women from across the country who had had horrendous experiences with breast implants. In the evenings, I pored over corporate documents, tracing which shipment of implants had been sent to specific plastic surgeons and carefully noting the health jurisdictions in which they operated.

I visited plastic surgeons' offices and informed them, as a

graduate student, that I was required to be scientific and value-free. I explained that I worked in an office with a stated feminist orientation and emphasized that my academic career as a social-work educator was at stake.

I asked them to openly and comprehensively share their perspectives, even using the phrase "man to man" at times, though I sometimes crossed my fingers behind my back. The plastic surgeons bought into this appeal.

I acquired a surprising amount of information, which I organized into flow charts that reconstructed the time frames during which various types of implants were unregulated.

I also met several fascinating women and further contacted others who called to question me about whether I recognized their voices or had any idea who they were married to?

Some called when they were either drunk or over-medicated, often breaking down in tears as they detailed how their husbands had discovered they had implants only during arguments about breastfeeding their infants.

I invited some of these women to my office to participate in groups and test a "living-room organization meeting strategy." In the process, I learned a few remarkable things from some very influential women.

Many of the calls were uniquely extraordinary, prompting me to pause, empathize, and reflect on my own feelings about coping with some disappointments and losses I had experienced in my own life.

One night after having a cup of tea with Stanley, who often worked late, I just returned to my office when I received a call.

"Mr. Gabriel?" the female voice said. "Yes?"

"A mutual friend of ours, Dr. Leenie Spressatoura, suggested I call."

"I'm not familiar with Dr. Spressatoura," I replied. "Let me just check my computer."

"Oh, you know Leenie."

"I'm sorry. I talk to lots of people, and I don't see her in my directory."

The woman offered a hint. "She's a sports psychologist working with the Olympic team of one of the former Soviet Republics."

"How interesting. If I'd met her, I would have documented our contact."

"I don't think so."

"Pardon?"

"Leenie is a friend of ours," she said, heavily emphasizing the word 'ours.'" Then she asked, "May I be familiar and call you Jeremy?"

"Certainly. And whom do I have the pleasure of speaking with?"

She briefly hesitated before answering. "I'm really sorry, Jeremy, I can't say, but if you figure it out, or, I should say, when you figure it out, let me know."

"Mila?"

"Good guess, sweetheart, but no cigar."

"Just checking," I said, sounding disappointed.

"You can be a real supercilious asshole, you know?"

"Have I offended you? I asked, genuinely apologetic. "Please let me know, specifically, and I'll try to remedy whatever."

I was cut off before I could finish.

"I have been in a lot of pain," she said. "None right now. I used to see Leenie professionally and, of course, I've followed her since she was a child."

"Does Dr. Spressatoura have implants?"

"No, I have. I called because I wanted to let you know how much I appreciated your information meeting, and the package you compiled was most helpful. If I can be of any assistance to you, I know lots of influential people in this old town, and I'd love to offer my assistance."

"Thank you very much!" I replied, still trying to recall which woman had spoken at which information meeting, "Have I asked you to respond to the 'Comprehensive Questionnaire'?"

"You have. I'm the one who only uses Buckley's Mixture."

There was something familiar about this caller, and I tried to recall who the voice belonged to. Could this be the cabinet minister's wife who, when anxious, insisted on using a heavy accent that was not the way she normally spoke?

"No names," she went on. "This is confidential, like alcohol treatments."

"And have I received the completed questionnaire?"

"No, no, no. I'm the one who said that if I had my implants removed and there were scars, I would decorate my chest with a tattoo, perhaps even a stick-on. I design them, you know."

"Oh, I remember!" I said, vaguely recalling an attractive, big-boned woman who was very well dressed.

"Hi!"

"Hi, Jeremy. How much do you remember?"

"I remember your smart outfit."

"Oh, you don't remember. Anyway, Leenie says to say hello."

"Well . . ., hello to Leenie, but I confess, for the life of me, I can't recall her."

"You can be such a big silly."

"Guilty! May my eyes drop out and my liver rot."

"I'm so sorry too. I really must go. Check the fridge. Leenie says you gave her a sweater from Paris."

"I don't think so."

"You are a self-important prick. You did so. I wasn't around when you gave it to her, but I saw her wearing it. It was cashmere."

"I can't recall ever giving a psychologist, "

She cut me off again. "I gotta go. Check the fridge. 'Bye.'"

I felt like I was verging on compassion fatigue. What a wingnut, I thought. I was living in Gatineau with Jon, a friend in the last stages of living with, rather than dying of, AIDS. It was a struggle to maintain the pace of my research and stay upbeat.

Earlier that day, my faculty advisor had leaned back in his chair and told me, "Breaking new ground on a vital aspect of the social construction of gendered identity and social rank is true scholarship. It's rare, and it's demanding, exhausting, even."

That evening, Stanley appeared at my door. "You're coming to my birthday party tomorrow?"

"I wouldn't miss it for the world."

When I got home, the message was waiting: Jon was back in the hospital. I stared at the note, the weight of it settling in. With interviews stacked for the next day and Stanley's celebration already set, I knew I wouldn't see Jon until six o'clock at the earliest.

By then, anything could have happened.

# CHAPTER 24

I didn't expect a birthday party to end with tears on a bus.

The day had started smoothly enough. Stanley, who had once served alongside Jon's grandfather, the late Bert Herridge, the longest-serving CCF/NDP member of Parliament from British Columbia, was in good spirits.

I arrived late to his office for the celebration and planned to stay only briefly. But news came that Jon was back in the hospital, and Stanley had already prepared something for him: an oversized crested card, carefully taped to a clear wrap over a matching plate. He handed me the package, along with a cake, and sent me on my way.

Cabs were nowhere to be found, so I caught the bus to the hospital. Somewhere along that jolting ride, I realized my cheeks were wet, silent tears slipping down as the city blurred past the window.

The Queen was scheduled to arrive the following day to unveil the statue of herself on horseback, a gift from the RCMP.

I worked that night and took a surprising number of calls.

One caller opening with, "Jeremy, have you checked the fridge?"

I recognized the voice. "Hi, how are you?"

"You didn't answer my question. Have you checked the

fridge?"

"No. Sorry. I go to the parliamentary café on the fifth floor whenever I need a break. Best Danishes in the country."

"Were you always such a stupid shit, or does it just come with the office?"

I hung up.

The next day, the House was aflutter with security and the hoopla of a royal visit. I had my special invitation to a select viewing area. Although I responded to and made calls to various federal departments, I was so excited that I could hardly concentrate on the tasks at hand.

Jon called three times, first, to tell me to look her straight in the eye when I curtseyed; next, to humbly bow my head when I curtseyed; and finally, "Under no circumstances, curtsey!"

I was authorized to leave the office at 1:30 p.m., as Her Majesty would be unveiling the statue at precisely 2:00 p.m. If I was not in place by 1:40 p.m., I could not be seated in the select viewing area.

At 1:15 p.m., the phone rang.

"Have you checked the fridge?"

"No."

"Are you going to apologize for hanging up on me? Is there no code of conduct for special assistants?"

"There may well be. I, however, am a policy analyst."

"And is not the essence of your vocation, excuse me,

profession, a healthy skepticism, a willingness to examine all truths, however marginal?"

"You've got me. Yes indeed. I'm very sorry I hung up."

"You just think I'm some muckamuck who could get you into trouble. That's why you sound so sincere, and you do it very well. I, too, am sorry l was condescending to you. After all, you don't know who I am."

"Apology accepted. How may I be of assistance?"

"Our friend, Dr. Leenie O'Malley, suggested I remind you she used to be your dancing partner."

"Now it's Dr. O'Malley and not Dr. Spressatoura whom I've now listed in my directory. But I haven't yet canvassed the Baltic Sea States Olympic teams."

"Don't lose it, Jeremy."

"Please," I replied, "I'm involved with some very serious research. Critical research."

"I know, I know. You're doing well, Jeremy. That's why Dr.

O'Malley and Dr. Sprezzatoura suggested we renew our acquaintance."

"Which Olympic team does Dr. O'Malley work with?"

"She doesn't work with an Olympic team at all; she works with the UN, focusing on internal refugees, people who have been tortured. She conducts diagnostic and therapeutic interviews to verify inland refugee status and gather evidence for tribunals aimed at bringing the barbarians to justice, though that rarely happens."

"I'm very sorry," I said, "but I have a rather important appointment."

"Yeah, yeah, don't I know it. Are you going to wear your granny's fox fur stole?"

"I don't have to take this."

"Don't hang up, Jeremy." The voice was now laughing, "This is historical!"

"You mean hysterical?"

"I mean, historical, didn't you wear your granny's fox fur stole when Princess Margaret came to town?"

"Who are you?" I asked impatiently.

"Here's a hint: You gave Dr. Sprezzatoura a cashmere sweater, and you used to be Dr. O'Malley's dancing partner, or at least you held the mirror for her while she played the piano and her grandmother danced the Irish Jig..."

A chill ran down my spine.

"Eileen Kathleen O'Malley, yes, I recall. Leenie."

"Good, now Dr. Sprezzatoura. Your mother gave her mother a cashmere sweater."

"Fine, but who are you?"

"Check the fucking fridge door, Jeremy. And get downstairs right away or you'll miss the Queen, you stupid bugger."

A parliamentary office generates tons of reports, several of which had been piled in front of the fridge since the last living room meeting. I dived at the small fridge and pried the door open. There was nothing extraordinary inside or on the door shelves, and I slammed it shut.

As I did, I saw it on the bottom corner: a bright pink stick- on Jackie decal, a poodle dog in a pink suit with mink trim and a pink pillbox hat.

I ran to the elevator. Normally, parliamentary elevators seem to travel at the speed of sound. No elevators. I ran down the hall toward the stairs only to hear the elevator bell. I ran back, jumping in just in time.

The lobby was deserted, except for the guards at the main door. I composed myself and walked to the Senate entrance foyer, where large portraits of various monarchs were displayed.

A well-dressed woman stood in front of the painting of Victoria and Albert. She saw me approach, pulled a magazine out of her purse, and ripped out a page.

"You want to order a Limited Edition?" Winona!

"Jesus, Winona. I almost thought it was Jake-Jack-John."

Then, partially shielded by a pillar, I recognized the big- boned woman from one of the living-room strategy meetings. She was dressed in a calf-length designer outfit and a two-toned stole.

On closer examination, it was two stoles.

"Stole anyone? Stole one of them for you, for old time's

sake."

Who is saying this to me? Who is this person?

Her gloved hand gestured to the enormous oil painting of Queen Victoria and Prince Albert.

"You know, they had weird genes. Most of their male

descendants had testicles that didn't descent. Edward, for instance, was infertile. Elizabeth II would have become Queen anyway, though she may have had to wait a while."

It was Jake-Jack-John!

I was speechless and in shock as he spoke again.

"I think I know why that little fascist shit who married the American hermaphrodite got along so well with Hitler. Together they were a couple of nuts."

"Hurry or we'll miss Her Majesty," Winona said. "C'mon, barrel ass it over, girls."

We ran through the rest of the Senate foyer to the front steps.

There we were, Winona, Jake-Jack-John, and I, standing among all these parliamentary guards who graciously let us go to the front of the group.

Her Majesty walked along the concourse in front of us. Jake-Jack-John quickly arranged one of the stoles over my shoulder and, equally as fast, Winona lifted it off and arranged it around herself.

The guards were curiously silent, and Jake-Jack-John made up for it.

As the queen walked in front of us, he waved a hankie and shouted, "Yea, Queen Elizabeth!"

Her Majesty looked directly at us and gave us an unmistakable, uncharacteristic sneer. Then she went to take the shroud off her horse statue.

What does one say after being seen up close by the Queen? I turned to Jake-Jack-John.

"I thought you were dead."

"Reports of my death were greatly exaggerated in an understated kind of way."

Winona explained. "Jake-Jack-John was a blue baby when he was born. He's like he is because of post-birth trauma, probably co-morbid with oxygen deprivation. When those thugs left him for dead, he was in such shock, he dissociated. That's how he survived. As the First Responders removed what they thought was his body, he opened his eyes, and do you know what he said?"

"I can't imagine."

"Get me to a doctor. I'm Shirley MacLean."

Jake-Jack-John broke in. "Then I passed out again and didn't wake up until I had several days and nights of near-death experiences. I soared toward the light and returned to the green light of recovery rooms, then soared toward celestial spheres again. Some of it was the drugs, but getting here has involved drifting in and out of therapies too numerous to list."

***

I finally completed my master's degree coursework. Due to some interesting circumstances, which, in the interests of national security, I'm both reluctant and unable to disclose, and, importantly, because I may choose to involve myself in similar endeavours on another occasion, the Honourable Benoit Bouchard upgraded the moratorium on silicone gel breast implants to an official national ban.

Shortly after we renewed our acquaintance in Ottawa, Winona was diagnosed with Alzheimer's Disease. Jake-Jack-John, who now styles herself as Jacinthe, recently married a former RCMP officer who attempted, unsuccessfully, to secure a nomination to run in the federal election as a candidate for the Reform Party.

The wedding was memorable. Most of the local people who came weren't sure which of Winona's girls had changed her name to Jacinthe. Nonetheless, Jacinthe was a beautiful bride.

I sat beside Winona during the ceremony, throughout most of which she sat, slack-jawed, staring into the near distance. There were moments, however, when she pinched my arm, nodded her head knowingly, and flirted with her eyes.

After the ceremony, when everyone else was having pictures taken, we walked through the graveyard and stood for a while by the grave of Dr. Peter. When we came back to the churchyard, some people remembered Winona in her prime and remarked that she once had been a "handsome woman, almost as pretty as the bride."

All five bridesmaids had once worn the uniform, municipal traffic RCMP officers who, somewhere along the way, developed testicular cancer. The cause was never officially confirmed, but many suspected it came from cradling radar guns in their laps.

Jacinthe had worked with them under a special contract in a classified division of the gender reassignment branch of the career-enhancement personnel department, guiding them through hormone treatments and vaginoplasty procedures.

Years later, Dr. Leenie Sprezzatoura and Dr. Leenie O'Malley updated me on their lives. Two of the former officers now call themselves lesbians and have fallen in love with each other.

Of the remaining three, one had settled into a quiet life in a Hutterite community in southern Saskatchewan, another was engaged to a career diplomat from a French-speaking European nation, and the last was pursuing postgraduate studies in social work in eastern Canada.

They had each gone their way, lives transformed beyond recognition.

But as I listened, a question lingered, unspoken.

What had really brought us all together in the first place? And was the story truly over?

# Chapter 25

## Other Graduate School Involvements

One of my social-policy professors, Peter Findlay, was also assigned as faculty advisor to my study partner, Lagong (pronounced *LaHong*), from Botswana. As well as seeing Peter in class, Lagong and I met with him individually in his office, where we discussed everything about our involvement in graduate school.

Peter was in his sixties, always well-dressed and well- groomed. He informed each of his mentees that he expected them to have clean fingernails in every encounter they had with him. Each term, he bought, in bulk, nail clippers and emery boards.

"I used to tell my students to trim their nails each Saturday night. Since none of you likely go to church on Sunday, I ask each of you to trim and buff your fingernails every Sunday night. The habit will serve you well in the professional world."

After a support session together with Lagong, we tested the ice on the Ottawa Canal.

"Peter suggested that I ask you to recommend a book that explains the socio-economic and historical-cultural ethos of the United States."

"Easy as pie," I assured him. "The Autobiography of Vanna White."

Lagong bought, read, enjoyed, and discussed the autobiography with me, then with Peter.

When I next met Peter, he said, "I was anticipating you would refer Legong to something like de Tocqueville's Democracy in America."

"You said within one cover," I replied, "and de Tocqueville's comes in two volumes.

"Well... that's an erudite demonstration of critical thinking. And looking at the reference in a different manner, it may well be enlightening to understand the pizzazz and hoopla of lumpen consumerism. Now I have a little mainstream assignment for you to test your literary elan."

I wondered what he was up to.

"On Saturday night, the local English literary society holds its annual poetry competition. This year's theme is 'An homage to deceased Canadian poets.' I'd like you to enter the competition, and I won't be satisfied with anything less than an honourable mention."

He went on to explain the competition format.

"Those vying for prizes," he said, "must recite an original poem they've written, one that is clearly an homage to a deceased Canadian poet."

He let that settle for a moment before continuing.

"Before you get to perform, each participant is asked by the panel of judges to recite a poem, a stanza or two if that's all you can manage, by a poet of the judges' choice. This may not necessarily be a Canadian poet. They're looking to see if you have any background in literature, specifically poetry."

"If you're unfamiliar with the poet they choose, you may ask to recite a sonnet by the Bard. And unless you're a Scot, that does not mean Burns." His eyes twinkled with amusement. "So… are you up to this?"

"I think so," I said. "Does Scots-Irish ancestry grant categorical eligibility for a Burns recitation?"

"Perhaps. It depends on how well you recite. Pay attention to tone and metre. Enunciate and convey expression appropriate to the text. If you succeed in the preliminary screening, you'll be asked to recite from a second poet, often an American."

I wasn't exactly flooded with inspiration. "I could do… 'A narrow Fellow in the Grass / Occasionally rides…'" I offered tentatively.

Peter ignored my suggestion and pressed on. "And if you pass this test," he continued, "you get to perform your own poem. Having an honourable mention will help when other professors and I advocate for a prestigious placement. It will allow me to say, 'He writes well.'"

"Also, when you're in the field, in any community, it's always of value to associate with the local Arts Council. And at any dinner party," he added with a faint smile, "it's handy to have at the ready a party piece to perform."

The night of the competition came, and I was a bundle of nerves, though not from stage fright or concern that I might forget the lines of my chosen poem. Tommy had taught me how to memorize; instead, it was the anxious anticipation of which poets the judges would ask me to quote.

I witnessed others not knowing a stanza from Elizabeth Barrett Browning or Robert Herrick, then eliminated in the first round because they couldn't quote, or poorly performed, a sonnet by the Bard.

When my turn came, I was introduced by name and as "from UBC."

The first judge repeated my name and said, "William Butler Yeats."

I was ready. "The Lake Isle of Innisfree," I announced. "I will arise and go now…"

The words came easily, my voice finding the steady, measured pace of the poem. When I reached, "*I hear it in the deep heart's core,*" the room seemed to still for a breath, then, to my surprise, a ripple of applause rose from the audience.

A second judge asked for Gregory Corso.

Before I left Vancouver for graduate school, a poet friend had taken me to a Gregory Corso performance. That memory was still fresh, and now I began with confidence.

"Anti-NAFTA Haiku, by Gregory Corso," I announced. The Mexican Zoo/has in ordinary pens/American cows.

The judge smiled, a trace of amusement in his eyes, and offered a polite round of applause.

"In the future," he said, "unless you are instructing an elementary class or performing at a seniors' facility, do not enumerate each syllable with your fingers. It's bad form. Your homage, Mr. Gabriel?"

I began with confidence.

"Fasting at the Fish Camp . . . Salmon and sage and smoke and salt brine/Sun and stars and fire always burning/Sacks full of fish from the river below/Pauline Johnson whispers me images..."

When I finished, the judges joined the audience in applause.

Their smiles and nods made me certain I'd at least earned an honorable mention.

Imagine my surprise when one of them leaned forward and asked, "Who is the Canadian poet?"

"I gave you a big hint," I replied. "'Pauline Johnson whispers me images...'"

A murmur of recognition rippled through the panel. Moments later, I was announced as the winner of the Anne Rice Book Prize.

When I came into his office on Monday afternoon, Peter rose from behind his desk, shook my hand, and wholeheartedly congratulated me.

He also invited me to a government/academic conference on the development of a Canadian social charter to constitutionally enshrine social rights.

\*\*\*

314

The school director invited six of us to meet with a sociologist from the UN High Commission on Refugees. The seminar aimed to explore how auto-ethnography can complement the broader processes of institutional ethnography in understanding and documenting the experiences of individuals seeking protective status from the Commission.

The dim, late-afternoon sun created a sombre mood only enhanced by the 1950s pole lamps with elegant, pleated shades. On the backs of the chairs were vibrantly woven saddle bags and camel bags, while prominently displayed on the wall was a large beige-and-grey Guernica poster.

With my notebook, pens, and genogram template in hand, I found myself recalling the opening lines of Henry Treece's poem *Guernica*:

In Guernica, the dead children were laid out on the sidewalk In their white starched dresses, their pitiful white dresses On their foreheads and chests, where the marks where death that came in like thunder

While they were playing their important summer games.

The guest lecturer, Dr. Noor Sobhani, was an elegant woman who wore her hair, beginning to grey at the temples, in an unassuming French roll. She wore a simple blouse of fine material, a dark, businesslike pantsuit, and a finely woven, multi-coloured, Kashmiri shawl.

Then there were her eyes, the most beautiful eyes I have ever seen, intense shades of blue and green under thin black brows. They conveyed a range of expressions, from sombre to empathic, animated to reflective. She nodded modestly when Bethany, the professor, introduced her many accomplishments.

She began by outlining the format of the presentation. After each participant had introduced themselves, the group discussed a "contract for social learning."

This included identifying any specific needs the seminar participants might have, such as accommodations for hearing, scheduled comfort breaks, environmental adjustments, or support for the emotional capacity required to process and document information of a potentially distressing nature, including content that could provoke anxiety or revulsion.

"Should anyone want to leave due to the content's traumatizing nature," she continued, "a breakout room is available, equipped with water, a desk, and writing materials. At your earliest convenience, your director and one of my assistants will join you there to debrief your perspectives."

She glanced around the room, her tone both firm and reassuring. "I ask that, if you must leave, please do so quietly. Once in the breakout room, sit at the desk and write your recollection of the traumatizing information in a triple-spaced format. This will allow you and us to analyze your perspectives further, using the additional lines for secondary and tertiary analysis."

No one left the room throughout the presentation. We learned about the importance of demonstrating respectful and attentive behaviour toward the interviewee. This individual, as a member of a racial, linguistic, cultural, or national group, may have experienced violations of their human or civil rights; both locally and globally, and might have witnessed crimes against humanity.

They also may have experienced imprisonment and torture, including sexual assault and genital mutilation."

She paused and made eye contact with everyone in the room.

"It is my understanding that your seminars typically take a different form; they're interactive, involving dialogue and debate."

She straightened a stack of notes on the lectern before speaking, her tone steady but deliberate.

"Whoever you interview will focus on specific factors, race, class, gender, sexual orientation, age, degree of ability, education, religion, language, dialect, and other identifiers. Are they a trade unionist, a journalist, a writer, or a cultural worker? What cultural elements, customs, taboos, diet, or attire might identify them, or their oppressors?"

Several participants shifted in their seats, pens poised over notebooks, the scratch of writing punctuating her words. One man leaned forward, elbows on the table, as though unwilling to miss a syllable.

"I'm now going to ask each of you to identify one or more of these factors, which I call 'collectivities,' from your particular standpoint. Your ability to consciously connect any of these identities with some aspect, whether essential or something you once thought incidental, could be critical in determining the extent to which they have been, and may in the future be, at risk."

She scanned the room slowly, making deliberate eye contact with each person. The air felt charged, the kind of silence that demands an answer.

"Give me something, each of you," she said at last. "How do you identify central or other relevant elements of your identities?"

A few people exchanged glances. Someone tapped their pen against the page, perhaps stalling for time. Others stared straight ahead, already deep in thought.

We took turns responding.

"Race; race and class; race and gender; gender and age, and, of course, race. Class and sexual orientation. Race, faith, location, and gender; gender and linguistic or cultural identity."

The Canadian Forces student said, "Rank, Social Insurance Number, year code. My adherence to conformity."

Absorbing it all, our guest lecturer continued, her voice even but deliberate.

"The process of an auto-ethnographic enquiry is foundational in determining the specific realities and intersections of the subject's multiple identities," she said. "If you inspire dialogue, you may ask, 'And do you know of others who have experienced similar consequences by any State or organization?'"

She paused to glance around the room, watching for any flicker of recognition before going on.

"If you receive no response, reframe the question: 'Do you know of others who have experienced consequences from the military?' Then ask about the police or religious institutions. Inquire whether there were any unofficial groups or individuals who caused or threatened various forms of harm."

Her tone shifted slightly, growing more methodical.

"In reference to the institutional ethnographic elements of the enquiry, you may have limited time to explore, identify, and document specific processes. Ask: What time of day were you picked up? Were you by yourself or with others? Did anyone witness this? Who? Where did it take place?"

The scratch of pens on paper filled the silence between her questions.

"What events first occurred on your way to the place of confinement? What details can you remember about your arrival there? Please share anything and everything that could help identify who played specific roles in your confinement.

"Were they in uniform, men, women, or both? Who seemed to be in charge, and how did they direct the activities of their subordinates? How long did you initially stay in this place? Were there windows or doors, and how many? What facilities were provided for basic bodily functions, and what was required of you to access them?

"Were you informed, either formally or informally, of the reasons for your detention? How was this communicated?

"When were you physically injured, and where did this occur? How long did this assault last? If assaults occurred repeatedly, who carried them out, individuals or groups? Were you offered medical attention and, if so, what was your experience?"

When she finally paused, the room seemed heavier. No one looked up from their notes.

Again, she surveyed the room. "How is this sitting with you?"

Surprisingly, there were no significant responses. A few of us stopped writing. The Canadian Forces student shrugged one shoulder and rotated his neck.

"Shall I continue?"

Silence accompanied a few nods in the affirmative. "At what point, if at all, were you given access to legal representation or any form of advocacy? Were you able to communicate with your family or associates? Did you feel discouraged or prevented from trusting anyone who might have been able to help you?"

"You escaped? Can you share more about this? Are you able to identify who assisted you? What concerns do you have about the safety of your family, friends, or associates?"

She went on, her gaze moving deliberately from face to face.

"Are there any barriers preventing them from seeking support from community or international agencies? If so, what are those barriers? Please consider mobility and access challenges related to race, class, gender, sexual orientation, age, ability, language, or dialect, as well as the risk that contact with colleagues or cultural groups might place on you, or them, in terms of detention and the indignities it involves. What cultural factors affect an individual's ability to seek assistance?"

The participants' pens moved quickly across their pages. "What else do you remember?" she asked after a pause.

Her questions came one after another, her voice calm but persistent. "Who else do you know who has endured experiences similar to what you have survived? Are you in contact with them, or with members of their families or associates?"

She shifted her notes, then added, "What are your current needs? Take time to signal through body language, pouring tea, or appearing to straighten your writing or recording materials to show the end of the interview. Pause and allow for silence. If the interviewee chooses to speak, document their response using either technology or written notes. Once they've finished, ask, *'What is your understanding of what will happen because of this interview?'*"

The room was quiet except for the soft scratching of pens and the occasional creak of a chair.

By the time the presentation concluded, darkness had fallen outside. We were all emotionally drained, reflecting on the responses to the clarifying questions we had asked.

As each participant surveyed the now grim and serious expressions on each other's face, it became evident that vicarious traumatization was a factor. It arose from delving into circumstances of individuals who had been brutalized, threatened, further traumatized, or disfigured.

During the question period, we inquired about standards of practice for interpreters and the availability of interviewers familiar with the language and customs of the detainee.

The presenter explained. "It is important to create space and provide the interviewee with the opportunity to question their interviewer or any translator about their familiarity with the interviewee's community. *Are they trustworthy?* Might the detainee, for familial, cultural, religious or political reasons, be reluctant to disclose information to the interviewer or in the presence of a translator?"

She asked whether we were aware of the CIA-funded experiments in depersonalization that had taken place at the Allen Institute in Montreal.

She then asked to reflect on the power and status imbalances between those designated as patients and the psychiatric personnel, along with their staff, who participated in this state-sanctioned institutional abuse.

We were asked to document any familial or personal experiences with military or paramilitary institutional abuse we, or members of our families or cultural groups, may have endured.

Additionally, we were instructed to examine the qualitative aspects of the abuse and explore its intersections with factors such as race, class, gender, sexual-orientation, ability, age, language, religious affiliation, and culturally defined rank within our families and communities.

We were also encouraged to define our standpoint and reflect on how this perspective influenced, or endangered, the negotiation of our survival.

# CHAPTER 26

**No Backup as a Setup**

Upon returning to my residence, I settled in to write, working steadily through most of the weekend.

*** 

The phone rang around noon. It was the police dispatcher. "A kid is stoned, and he's going to kill himself," she said without preamble. "We can't get hold of his mental-health worker, and the two constables there need someone who isn't in uniform to talk the kid down."

"Do you have any other information?" I asked. "Just that they want you there right away."

My car was in the shop, so I grabbed a clipboard and jumped on my bike, pedalling hard through the damp streets to a low-end hotel only a few blocks away. I arrived within five minutes of the call.

On the third floor, the hallway smelled faintly of stale smoke and fried food. Two uniformed officers stood outside a closed door, their expressions tight. From inside came the muffled sound of crying, punctuated by sudden thuds against the wall.

"The guy is drunk or stoned, or both," the younger officer said. "He claims to have a knife and told us he was going to kill someone."

"Anyone specific?"

Noises continued in the room. Ignoring my question, the older officer added, "The hotel has complained."

His colleague said, "He has a mental-health worker, but we can't reach anyone. Your job is to talk him down. If you need anything, we'll be right outside the door. Take the time you need."

"Give me a minute."

I walked down the hallway to poise myself for whatever I might encounter behind the door.

"Where are you going?" the older officer asked aggressively.

"As part of my training, I need to compose myself before beginning an intervention."

Sooner than I would have preferred, I turned and walked to the hotel room door and, standing to the side, I opened it. The officers stood back on either side as I surveyed the scene.

A blond male, in his early to mid-twenties, stood on an old-fashioned wall radiator by one of the two windows in the room. Another door led to the bathroom. The young man had his back to me, one hand resting on the window.

"Are you in here alone?"

"Not unless you're God."

I looked over my shoulder as one of the officers whispered. "You'll be okay. We'll be right here."

I moved slowly into the room, leaving the door open. "I'm not God…"

"Well, close the fucking door. Can't you see I'm just in my gouaches!" The youth began to cry, his heavy sobs shaking him so intensely I feared he might lose his balance.

I looked at the officers and nodded to the wall between the hallway and bathroom doors.

"I'm going to sit down by the wall," I said. "Anything you want to tell me will be okay. Do you have a brother or a good friend?

You're going to be all right. Talk to me. You're not going to blow my mind, and nothing you say will shock me. I can accept anything you tell me."

I closed the entry door, then sat with my back against the wall. If I were attacked, I could defend myself by holding off the youth with my feet until the officers entered the room.

He began talking, his words tumbling out in a rush, the effects of more than two hits of LSD clouding his voice. He wasn't sure exactly how many, only that he had been smoking pot and drinking heavily.

"Hard liquor," he added, still balanced precariously on the radiator. "I caught my girlfriend and her boss having sex, and I want to kill them. And maybe myself. I don't know where to start. If I don't kill someone, I'll have to go to work tonight. This has happened before. She isn't the first girl I've picked up, cleaned up, bought nice clothes for, paid to get her hair done, taught her how to enjoy sex with me. I just can't cope with this happening again, and again, and again."

His voice cracked, and then the sobbing started, deep, shuddering waves that shook his whole body. I stayed silent, watching him, waiting until he could find his voice again.

"Nice girls and virgins just see what I've got between my legs, then they run out of the room. I might as well turn gay."

He was employed at a local shopping mall and worked from 4:00 a.m. until 8:00 a.m. and from 9:00 p.m. until midnight. He also claimed to "have an erection that would choke an ostrich."

"I might as well jump out of the window and kill myself right here and now."

"No, don't do that." I said, "We're only on the third floor. You may seriously injure yourself and find yourself paralyzed. Then if you want to kill yourself, you won't be able to do that, and you'll curse me and everyone forever for not stopping you. Please come down. If you've ever had a brother or a good friend, let me be that person for you now. We can solve this."

Slowly, covering his crotch with both hands, he jumped onto the grimy carpet by the bed.

"This will go better if you sit on the bed," I said. "Then you can tell me what this is all about."

"I already fucking told you!"

"I'm still listening, and I'm sure there's more."

He arranged himself on the bed, as far away from the window as he could get, close to where I still sat on the floor. He soon started talking.

As a child in elementary school, he had been prescribed growth hormones that made his penis "grow to the size of a club."

He had not attended junior or senior secondary school but had been educated at home by *"an in-home tutor."* He rose from the bed and sat amid the scattered clothes and towels on a lounge chair.

Drawing his knees up in front of himself, he made an unsuccessful attempt to conceal his rather prominent erection.

He was twenty-one years old and had an eight-year-old son "in Alberta by the in-home tutor. I'm coming down from the LSD and need to smoke some pot to ease myself down, but my hands are too shaky. I don't know if I have papers, I'm too fucked up to look for the pot if there's any left and if there is, there's cops out there."

We talked more, and I encouraged him to take a shower. He gathered his towels and agreed to accompany me to the local hospital to obtain care, possibly medication to bring him down.

While he was in the shower, I checked the hallway, but the police were no longer there. I ran down to the front desk and asked why they had disappeared. The desk clerk started to provide directions to the local detachment.

"You just go over to the Mile 0 sign, then turn right and go up a couple of blocks, big white building, flagpole outside."

I was furious and raised my social-worker ID. "The police were here and called me to an incident on the third floor."

"Oh, they left shortly after I began my shift, maybe five minutes after 1:00 p.m."

"Are they even in the building? Or in the coffee shop?"

"Might be. But the staff have probably cleared away the buffet."

I hurried to the ground-floor restaurant, scanning every table, but saw nothing. Back at the front desk, I asked to use the phone.

When someone finally picked up at the detachment, I identified myself and my location.

"I need an unmarked car at the Lone Star front entrance. Wait there until I get down with the patient; he needs to go to the hospital."

"The detachment is in union–management negotiations with the city out at Guillam Lake," the voice replied, "and we don't have any unmarked cars to spare. Hold the line, Jeremy."

I stayed on the line, my eyes drifting to the stairwell as I imagined what the young man might do if left alone too long. The thought sat like a stone in my gut. Then I remembered, there was no union acting as a bargaining agent with the city.

If they thought that line would make me back off, they didn't know me. I was already thinking of how I could challenge it, if it meant convincing them to send me the car I needed.

The dispatcher came back on the line, laughing. "I've checked, Jeremy, and I'm supposed to tell you he's in your hands."

There was more laughter from others heard in the background.

"I want to speak with the watch commander. Stat!" The dispatcher promptly hung up.

I ran back upstairs and opened the door. The youth wore black pants and a white shirt. He looked like a young Mennonite pastor, the impression vanishing when he asked, "Do you think I'll be fucked up forever?"

I persuaded him to walk with me and my bicycle to the local hospital. As we passed the library, he asked what "that new-fangled building" was.

"The library."

"Does it have a librarian?"

"It sure does. Two. And all kinds of books, magazines, and videos."

"Well, one of those librarians better have a neck like an ostrich."

He jumped into the middle of the road, narrowly missing a car, and proceeded to perform a highly sexualized dance, thrusting his groin at the passing cars and people walking their dogs along the footpath.

When we reached the hospital, we walked through the main entrance. Showing my ID, I told the admitting clerk I had accompanied a young man with a "non-organic brain injury who needs to be assessed as soon as possible."

"I know who you are," the admitting clerk said, "and you know better than to show up at Admitting. You can walk around the building and enter through the Emergency exit. But he sure doesn't look to me like he's been injured. Now skedaddle!"

As we walked down the ramp to Emergency, a nurse taped a

hastily drawn sign on the glass entrance: EMERGENCY CLOSED UNTIL 3:00 P.M.

I felt devastated. What was I going to do? It was already enough trouble getting the young man to walk with me to the hospital.

When in doubt, use food. Even Freud fed his patients.

"You're probably hungry," I said. "Come with me, and we'll get some lunch. It will do you good."

We walked the bike around to the George Dawson Hotel, known for having the best Sunday buffet in town. I began each day there with a hearty breakfast. As we entered, I said to the server, "I'll have the usual. Please bring coffee and a menu for my friend."

The restaurant was crowded with tables of large families, many of whom had just attended one of the several evangelical congregations in the community.

Their glaring stares, grimaces, and whispered comments clearly conveyed an unwelcoming tone.

Here I was, still in my Sunday-best blue Oxford cloth button-down shirt, sitting next to a tall, handsome, young man who looked as though he'd just stepped out of the shower.

His shirt was open at the neck with two or three buttons undone, giving the impression that he'd just come from an audition or performance at a male stripper event.

He ordered a clubhouse and fries, and when the server came back, he said, "I've got to get the fuck out of here. Bad, bad energy here."

"Take my bike," I said, "Be back at the hospital by three o'clock."

I sat, stunned, documenting the hotel experience on my clipboard. As the evangelicals got up to leave or pay the server, some of the patriarchs, and a few of the women, looked directly at me and shook their heads.

One woman, her fork paused midair, turned toward us with narrowed eyes.

"In a public place, on a Sunday," she said, her voice sharp enough to cut through the low hum of conversation. "May God have mercy on your soul, if you've got one."

Her tablemates nodded, and the murmur of disapproval spread like ripples in still water.

When I met him later at the Emergency entrance, he had propped my bike behind some rose bushes.

"You've got to watch out for people in this town," he said. "They may steal your fancy bike."

We were seen immediately by a doctor whose father had been a physician in town. He displayed open, non-judgemental, professional behaviour as his young patient revisited the ingredients of his recent impairment and his serial unfulfilling experiences with women.

"I'm tired of cleaning up lummoxes only for them to dink their bosses or their brothers or cousins. There are real trash hillbillies in these parts."

Accompanying this judgmental perspective, he burst into tears and said, "I'm still really stoned. I'm coming down fast, but I'm still blitzed."

The doctor agreed to admit him and prescribe something to bring him down. He said he'd provide a medical certificate, allowing him a sick day from work.

"Substance abuse, particularly poly-substance use involving acid, pot, and alcohol, is a medical problem that can aggravate existing social and mental-health stresses. You'll feel better by tomorrow morning, and then we can talk about support options, including appropriate and non-judgemental counselling."

The young man described his car, and the emergency room doctor wrote a pass to allow me to park it in the doctor's lot.

As I left for the hotel to retrieve the car, the doctor said, "I'll see you after you've parked."

When I returned, he asked, "Do you have any wine at your house?"

"I do."

"Let's go to your place so you can tell me everything you remember about being called to this situation."

# CHAPTER 27

When I arrived at the office the next morning, I filled in an overtime form, noting the time of the call and when I got home. I didn't document the time I spent over a couple of glasses of wine shared with the doctor.

Barkie, my clinical supervisor, immediately returned the form, saying.

"I'm denying this. You are sanctioned to work with children, youth, and their families. It's a privilege that this is your mandate. I've checked with the hospital, and the young man you spent five hours with is twenty-one years old. Overtime denied."

I promptly called the union and got through to Kim Smith, the business agent.

"Jeremy," she said without preamble, "they didn't call you out as 'Jeremy, nice guy.' The RCMP called you because the government employs you as a professional social worker. You're the chief shop steward, the local chair, and a member of the component executive.

"Call the regional manager. Tell them you want overtime, call-out time and double time since it's Sunday. And emphasize that you want the regional manager to make an official complaint."

Her tone softened slightly, though the words were still brisk. "As you know from steward's training, I normally don't suggest someone act as their own steward. But this is the exception that proves the rule. You may not want to talk about this with too many people. Confidentiality and all, you know."

After this call, and before I called the regional manager, the young man's mother called from Alberta to thank me for saving her son's life.

"He's done this kind of stuff before. It all goes back to when he was prescribed growth hormones, and then he was sexually assaulted as a twelve-year-old by his tutor."

"Were there charges?"

"We signed a non-disclosure agreement and will have to pay back the settlement if we discuss any of this other than for diagnostic and therapeutic purposes. I guess that covers me talking with you."

I called the regional manager to grieve the denial of overtime.

"Just take a couple of days off," he advised. "Nobody will say anything, and you'll be paid for the days off."

"I don't want to take any time off. I have court prep and court this week. I'm also doing an adoption as the relinquishing parents' worker. I've even got consents from the putative father, so I can't reschedule any of this." I replied.

"Take a couple of days off. That's how this kind of thing is handled. Nobody will say anything." He sighed.

"What do you mean, 'Nobody will say anything'? I want you to make an official complaint. In fact, I've been advised by the union to ask you to make an official complaint to the RCMP. Let me know when you've talked to the watch commander."

"Just take the days off. Do I have to spell it out in writing?"

"I'm missing something here. I want documentation on this incident. If I choose to take 'compensatory time off', I'll do it in accordance with the master agreement."

"Do I really have to spell it out to you?"

"Yes, spell it out to me."

"You were in a hotel room for an indeterminate amount of time, and when you entered, witnesses noted that the subject mentioned he was in his underwear. Take a few days off."

I hung up the phone.

Years later, after the seminar with the UN sociologist, I found myself writing furiously. Her words had stirred something I had carried for too long. As I sat at my desk, I tried to make sense of it all, the incidents, the patterns, the silences. My response began to take shape on the page:

Within the limits of this assignment, I want to examine two incidents. Incident One, the day I first learned back-up would never be provided. And Incident Two, the hotel incident. Together, they reveal themselves as expressions of state paramilitary homophobia.

Even as I wrote the words, my supervisor's warning echoed in my mind: The RCMP are not going to provide you with back-up. At the time, I thought it was neglect. In hindsight, I saw the outline of a strategy, carefully staged.

The young man in the hotel room, stoned, broken, trembling, was not the real danger. He was the excuse. His crisis provided the perfect stage for something far more deliberate: leaving me alone, exposed, vulnerable.

Structural preconditions were put in place, I wrote, to set me up for some form of undetermined, and possibly violent, assault. It sounded cold on the page, but the truth was anything but. The police had tools they refused to use. They could have sanctioned him under the Mental Health Act. They could have acted. Instead, they laughed. A dispatcher had even joked: I'm supposed to tell you he's in your hands. Ho ho ho.

The more I examined it, the more I realized how deep the rot went. My identity had become their precondition. My being gay had been turned into a threat, something they could use to justify risk, to build a strategy around. It wasn't enough to isolate me; they needed to make sure no one could ever be seen defending me, no one could ever risk being associated with me.

It turned into locker-room talk. Jokes about soap dropped in showers. Quips that if I were there, someone would end up pregnant. They were meant to humiliate, yes, but also to control. And if anyone tried to stand by me, they too became targets of mockery.

It was a system, I realized, that devalued me until I was expendable. I could be called into any situation, no matter the risk, and they could walk away. The jokes, the sneers, the constant reminders of my difference were not random; they were the background music to a much darker plan.

I wrote on: Inevitably, the process devalued me to the point where I deserved whatever happened. If they could find a situation requiring back-up, and a social worker could be placed there, it would be me. And if possible, the police could disappear.

Even as I put the words down, the anger returned, but so did the clarity. What had once seemed like isolated slights and casual cruelty now revealed itself for what it was: structural, calculated, and designed to make me afraid.

Curiously, when working together on high-profile sexual abuse investigations, we had no problems. Some talked about modelling your skills. 'If I go to court on this one, you'll spend time with me on the phone? Great, buddy.'

Driving home from an arrest, the silence in the cruiser was broken by the constable beside me. His voice carried the sing-song lilt of someone who had practiced this kind of needling before.

"You and your women colleagues are close to one another, hey? Let me know when Sarah gets her period, hey? Hey, I'm speaking to you."

I clenched my jaw, staring out the passenger-side window as the snow rushed past in the beam of the headlights. "Get off my fucking back," I muttered.

The constable smirked, leaning a little closer to the wheel. "Yeah. You want me on your fucking back. Let's put this on the radio. Jeremy has finally popped the question to me. What say we call all the guys and the truckers? You'd like that, wouldn't you?"

My temper flared. "Stop the car!"

He ignored me, his eyes fixed on the empty stretch of road ahead.

"Stop," I yelled, this time louder, my voice echoing in the small space of the cruiser. "I want to get out."

The constable only chuckled. "It's forty degrees below zero out there. You'll freeze to death."

"A better fate than riding with you," I shot back.

He finally turned his head toward me, his expression unreadable. "Look, Jeremy. I'm gay myself."

I barked a bitter laugh. "Well, you have a bloody weird way of showing it."

"I'm just trying to toughen you up," he said, his tone defensive now, though he kept his eyes on the road.

The car grew heavy with silence after that, the heater humming, our breath misting faintly in the air. After a while, I asked quietly, "How do you survive?"

He hesitated before answering. "I live with a woman."

"She knows?"

"She's a lesbian. It works out well for both of us."

I turned to look at him fully, my voice steady. "Do you know the expression, 'You protest too much'?"

"No," he said with mock curiosity. "Enlighten me."

"You want me to enlighten you?" I leaned forward, speaking evenly, as though lecturing a class. "It's originally a remark by Oscar Wilde about gay men who put down other gay men so no one will suspect they're gay. There's also a psychiatric term for gay men who do that, it's called reaction formation. It's Freudian. It's in the same category as those Jewish men who used to wear cast-off Nazi uniforms and goose-step through the camps, pointing out those who should be put in the ovens."

"That's a little extreme," he muttered, his knuckles tightening around the wheel.

"So is what you do when you pick on me," I snapped. "Another perceptive social scientist called it 'the banality of evil.' That's what your homophobic harassment is: the banality of evil. I couldn't live like you and your woman friend live. I want to change the world. I want to make the places we work safe for everyone."

We drove the rest of the way in silence. When he finally pulled up outside my place, he let the engine idle and turned to me with a sly grin.

"Can I come in?" he asked.

"I don't think so," I said flatly.

"You'd enjoy it."

"I don't think so."

"Why not?"

"Because you're in a relationship, such as it is and for me, any intimacy with you would be like adultery."

340

His smirk faltered, but he said nothing more. I got out and shut the door behind me, grateful for the cold air.

The only people I could talk to about this were two deputy sheriffs in my union local. When I recounted the incident, they gave me the same blunt advice, their faces set with grim experience.

"Don't fuck with them," one said. "They can be real assholes, not just your garden-variety pricks with asshole tendencies. Real mean assholes."

If I hadn't been able to approach this work with humility and a deep awareness of the privilege it entails, if I hadn't loved it enough to deconstruct my identity while remaining genuine and supportive to women, children, and youth, I would never have survived in this profession.

Drawing on my background in theatre, my training in conflict resolution, and the discipline of clear, direct communication, *"No, don't do that. We're only on the third floor."*

I managed to pull him back from the edge. Without those skills, the morning could easily have ended in tragedy, with headlines screaming across the papers:

*Homosexual Social Worker found Dead in the Lone Star Hotel: Police Issue Statement Indicating "Fowl (sic.) Play Not Suspected.*

*The local paper might even have added:*

*Follow-up with the Department of Agriculture confirmed that no foul play appears to be involved. Animal-welfare personnel were not called to the scene, and the Department of Agriculture will be making no further comment.*

After I turned in the assignment, I got a call at my dorm from Dr. Bethany Lazarus, the professor from my Human Behaviour in the Structural Context class.

"Did you make an official complaint?"

"I think the regional manager did."

"I'd like you to follow up on this and give me an answer, yes or no, and I'll give you an A+."

On a Friday afternoon at around 4:00 p.m., I called a number in Ottawa that had been given to me by a high-school friend who was a high-level RCMP officer. Having taken intake calls myself, I had spent many Mondays handling intake solidly for the better part of five years, until I eventually found myself on Friday intake.

Both days come with their own labour-intensive and emotionally demanding challenges. That afternoon, during my call, I asked clarifying questions about the process, such as what kind of evidence was required and what to do if I didn't know the names of the specific officers involved.

I spoke with the intake officer for over an hour. He confirmed that no official complaint had been filed and asked if I wished to make one.

After obtaining the necessary information for Bethany, I paused to reflect: Do I want to revisit this period of my life? What might be the consequences of doing so while being evaluated for a top-level security pass for an internship in the House of Commons?

The supportive officer said, "These members may still be engaging in this intolerable and possibly illegal activity."

That convinced me. Someone intelligent and empathic within the head office of the RCMP felt encouraging.

On Monday, I told Bethany, "No, there wasn't an official complaint made, but there is now."

"Then know that at any time in this process, I will be there to support you," she replied. "Making a complaint to any agency with control over human lives involves experiential learning. You will be a much more supportive social worker or policy analyst for your clientele, perhaps whole categories of people, if you go through this process."

The process was, to use a social-work term, "crazy making." Bethany and her German Shepard, Woof, accompanied me to an interview on the fifth floor of the CSIS building on O'Connor Street.

Two Francophone officers introduced themselves and explained that they had been assigned to this case because each had a son about to "graduate from Depot in Regina."

They mentioned that this assignment would enable them to afford attending both the graduation ceremony and the weddings, as both sons were set to marry "Saskatchewan girls."

"What do the girls do?"

"Teacher."

"Nurse."

Later in this interview, while summarizing the evaluation meeting I had had with the human resources executive from the BC Public Service, I recalled his mentioning that two officers had visited.

At the time, I asked, "Did you diarize the meeting? Were arrangements made in advance? By phone? Did anyone take notes?" I was told I was taking this too seriously.

I then delved into details of the discussion, specifically the moment when I was asked if I was a homosexual.

"And did you tell him?" one of the officers asked.

"More or less."

"You told him you are homosexual?"

"Yes."

"Why did you do this?"

"Because not only is it a prohibited article of discrimination by the Canadian Association of Social Workers, but I'm also not embarrassed about my identity. And at this point, I would like to remind you that at the beginning of this interview, each of you told me you were heterosexual."

Looking at each other and feigning shock, one of the officers said, "We said no such thing. This interview is being taped."

"Not only that, you told me that you both are fathers, that your sons are also heterosexuals, and that they plan to marry girls while you're in Depot en route."

The same officer answered for both. "Surely you're not saying that our being fathers can in any way relate to your status as a homosexual in northern British Columbia?"

"It contrasts with my identity as being gay in northern British Columbia, here in Ottawa, in this room, in this interview."

They nodded in unison, and one of them even gave me a reassuring smile.

"We would like to emphasize," he said evenly, "that we'll be value-free in all our enquiries."

Trying to break the stiffness in the room, I let out a short laugh. "That's what I'm worried about."

Soon afterward, my high-level security pass was approved, and I was photographed for my Parliamentary staff identity card. The formality of it all left me both proud and wary, knowing the stakes behind the work.

As well as thoroughly enjoying working alongside Joy, I confided in her about the official complaint and the unsettling circumstances surrounding it.

"Good," she said without hesitation. "Get the bastards. If I can be helpful in any way, I'm always willing to get on the phone."

Joy was, without question, the best boss I'd ever had. Together, we strategized carefully, poring over the results of an access-to-information request that had landed on our desks. The files detailed regulatory data on various corporate designs of silicone gel breast implants and their approval processes.

Her tone sharpened as she pushed the next step into my lap. "Your task is to locate the documents that were excluded from the evidence dump."

One day, bright and early, dressed in a crisp grey linen shirt, my signature blue Oxford cloth button-down, and a coordinating light blue tie patterned with white Scottie dogs, I arrived five minutes early for my appointment with the acting director of the Radiation Protection Branch at Health Canada, where medical devices were assessed for regulation.

Surprisingly, the acting director looked young and was as well-groomed as I was. I gratefully accepted his offer of coffee.

"Do I get to wear a dosimeter while I'm here?" I asked.

"How do you know about dosimeters?"

"Basic health and safety training, including yearly and lifetime acceptable body burden levels, that sort of thing." With a laugh, I added, "I'm your typical British Columbian roentgenologist."

"And you learned this in BC Public Service?"

"I learned it from a scientist friend, Dr. Rosalie Bertell."

"You're a friend of Sister Rosalie Bertell?"

"Rosalie has visited my home several times for dinners and evening slide-tape educational presentations on radiological epidemiology. We were both members of the Canadian Coalition for Nuclear Responsibility.'

"Yes . . . CCNR," he replied. "Dr. Edwards, Dorothy Golden

Rosenberg, and the rest of his informed crowd. The bane of our lives here."

"The reason I'm here is to ask some general overview questions about the medical devices section. As a graduate student with a professor who insists on thoroughness, I'm starting with you to gain a comprehensive perspective. That way, when my professor questions me, I won't come across as superficial or as though I've cherry-picked only the information I need."

He raised his brows. "And what is that? I presume you do need to know our policies on dosimeters, when we wear them, and how often they're checked. We do very little experimental work here, and that work is largely contracted out and written up in peer- reviewed journals. That is, in large part, a protection against anyone in the department getting too much insight into specific industrial developments, then taking competitor information to other corporate entities."

"And the specific information, does the research conducted on medical devices largely get contracted out as well?"

"Same principles apply," he said. "As civil servants, we ensure that we aren't in a position where we could entertain fantasies of windfall profits by engaging, or even appearing to engage in corporate espionage."

"So your work is largely with the Control Board?"

"No, they run their own show. We conduct random audits across various operations of AECL, Atomic Energy of Canada Limited. Budget-wise, in terms of a ballpark comparison, the medical devices department isn't a major priority for the Branch; it accounts for less than 2 percent of our work."

"For that small a percentage," I said, "they seem to generate a fair bit of paperwork."

"You're referring to your recent FOI request?"

"Yes."

Sitting across his desk, tenting his hands, I noticed that this well-groomed, handsome, intelligent man was not wearing any rings. The thought crossed my mind that he might be another RPQ, a Rich and Powerful Queen. I continued.

"We received a superabundance of information, so much, in fact, that it is almost beyond the capacity of our office, even with secondments from our talented research unit, to adequately assess the info."

"And?"

"I'd like to fully understand how decisions are made regarding the types of data that are exempted from an FOI request."

"Usually, information that may relate to corporate security."

"But shouldn't a Member of Parliament have access to such information?"

"Theoretically, yes. Practically, no. We're specifically mandated, through delegated authorities in legislation and regulations, to ensure the integrity of corporate interests. Bear in mind, we primarily work with AECL."

"I'd like to understand better, particularly with respect to the medical devices branch, is there any form of appeal process to challenge the decision to exempt categories of information?"

"Interesting question, but regretfully, no. We have strict guidelines to ensure the security of corporate information and any research and development conducted by our private sector contractors."

"Are there extraordinary circumstances where information not covered by corporate issues may be exempt from a request?"

"Certainly. This would take the form of a special request from the office of a cabinet minister if information released could conceivably embarrass the government."

"Could you give me an example?"

"I'd very much like to do so, but my current role as acting director constrains me. Eventually, I would like to become the director, and that may be possible if the current director's secondment to the International Atomic Energy Agency evolves into full-time employment. Discretion and the need to respect the intellectual property of corporate interests prevent scientists in my position from being forthright in our opinions on public entitlements to specific information. By the way, have we met one another before?"

"Do you attend the coffee klatch after the eleven o'clock service at the Anglican Cathedral?"

"No, but were you at the Royal Society of Canada Conference

on Nuclear Issues in Vancouver?"

"The CONIC Conference?"

"Yes, the conference on nuclear issues in the community." There was a distinct silence as we sat, evaluating one another,

him re-visiting the information he had provided, and me pondering whether he recalled a voice behind a mask of radiation-yellow plastic switch plates, partially obscured by a monk's caul. Over this, I had worn a red velvet smoking jacket turned inside out while addressing the conference floor microphone.

Then he smiled and said, "Well, you can tell your professor we got to *'Yes.'* Isn't that one of the goals of The Seven Habits of Highly Effective People?"

"Rosalie would say there were ten." We both laughed.

"Sister Rosalie definitely would. Always comprehensive, her Tri-State leukemia survey examined ten million subjects, each with the actual body burden of radiation they received during their treatment. The study was widely praised for its statistical comprehensiveness."

"Thirteen million," I replied.

"There's one thing you might be able to do for me. My boss has asked me to locate a statement made by Richard Nixon in which he mentioned authorizing a bonfire of chemical and biological weapons. Could you help me find anything on this?"

After a generally convivial meeting, this felt like a test. He could easily have asked his staff or an intern to search the poli-sci library or the peace research collection. Instead of suggesting that, I simply said, "I'm on it."

# CHAPTER 28

**Nobel Prize-winning**

Stanley asked me how the meeting went. When I mentioned the request about Nixon's bonfire remark, he said, "Give me a minute. In less than one minute, he had found a Scientific American publication of articles by Nobel Prize-winning authors.

"Here it is, in a 1974 article by Alva Myrdal entitled 'The International Control of Disarmament.' I knew I'd read this information. Listen to this: 'In the early days of the Geneva disarmament conference, there was often talk of such bonfires of bombers and other weapons. In the convention prohibiting the production of biological weapons, the destruction of the existing stockpiles is explicitly required. President Nixon, in 1969, made the magnanimous gesture of promising such destruction of U.S. stockpiles . . .'"

Back at the office, I realized the evidence dump was one reason I didn't immediately notice the decal Jacinthe had stuck on the office fridge during the living-room meeting. Using my institutional ethnographic skills, described later by a highly placed informant as "his forthright charm," can you guess who I called with the information Stanley had provided about Richard Nixon? In the end, I obtained two pieces of critical information:

A specific breast implant was designated by the US Food and Drug Administration as "not to be implanted in women in the

352

continental USA." This restriction allowed them to be exported to Canada, Hawaii, and the Philippines, and they also could be "implanted" into RCMP officers undergoing gender reassignment at the Mayo Clinic.

The other piece of information proved critical to elevating the national moratorium to an official ban. The individual with "sole distributor rights" to import the toxic technologies acted as the bagman for the prime minister and five Quebec cabinet ministers.

We strategically introduced a carefully crafted question during Question Period, just before the House prorogued for the summer. This would have left the damning query on the Order Papers for all interested parties, including journalists and opposition Members of Parliament. Within two days, a meeting was arranged with the ministerial staff for the responsible cabinet minister.

An aged, overweight, Quebecois ministerial assistant and political appointee asked, in a supercilious yet sarcastic tone, "Why are you involved in this issue? The matter exclusively concerns women."

"Not exclusively."

"Really? Please enlighten us."

"I'd love to. Your perspective neglects to account for those members of the transgender community who seek augmentation as part of their gender reassignment."

"I don't understand. At all."

I was pleased to share the details. "For instance, members of the

RCMP lack protection from any form of union. They also receive inadequate occupational safety and health training and equipment, both during their training and in the field. Often, they place radar-tracking devices on their laps without wearing protective lead aprons. As a result, some are later diagnosed with testicular cancer. A smaller percentage of these officers opt for gender reassignment, and among them, some already have had toxic implants placed subcutaneously."

"For any one of a number of reasons," he said, "I am wholly aghast that this can occur."

"This procedure usually occurs while the patient is undergoing hormone treatment and vaginoplasty surgery."

A young parliamentary assistant had to translate.

Within a matter of days, the moratorium was upgraded to a national ban. We achieved everything else we aimed for. Notably, we initiated dialogue on the compensation of provincial and territorial health ministries by large corporations. Additionally, we developed a system for regional centres to store explanted implants in forensic blocks, facilitating both affected individuals' litigation and, equally important, treatment assessments.

Joy had been at the International Labour Organization (ILO) in Geneva, and I met her at the plane with a bouquet of roses from the garden of my friend David's mother. David and I had been housemates during organized nonviolent demonstrations against the Trident submarine base in Bangor, Washington.

Along with Greg and Taeko, we managed Earth Embassy, a

clearing house for a hot-spot map of environmental issues, most notably, the locations of community groups that intervened in the Bates Royal Commission of Inquiry into Uranium Mining.

After successfully securing a ban on uranium mining in British Columbia, Taeko returned to Japan. Meanwhile, David moved to Ottawa to serve as the senior research assistant to Jim Fulton, the New Democrat environmental parliamentary critic and an alumnus of the UBC School of Social Work.

On the way back from the Ottawa airport, Joy went through her mail and found an elaborate invitation from the government of Kuwait to attend a reception at the Museum of Civilization, thanking Canada for its support in the first Iraq War.

"I'm not going to this! It's a bloody spoils of war event."

"I'll go," I offered.

"Sure, go ahead, you deserve a free meal."

David's boss, Jim, had also received an invitation, and I asked Stanley if he was going. Neither he nor Jim wanted to attend, so I had three tickets.

Throughout the academic year, David and I had coparented his daughter, Manna, along with her friend, nine-year-old Yuka, the daughter of Japanese feminist rock musician Tomiko Ichihara. In Japan, after a strong political reaction to Tomiko's performance of Canadian folk singer Bob Bosson's satirical song, "Mayor of Pacifica," she felt it necessary to get Yuka out of the country.

One of the greatest honours of my life was co-parenting

Manna and Yuka, an "emancipated minor."

Known as "Jeremy's nieces" by members of the Black Women's Feminist Social Work Caucus, the girls also came to know numerous people in an Ottawa Anglican congregation. We attended Sunday services to help Yuka learn the diction of the Anglican Book of Common Prayer and its successor, the Book of Alternative Services.

As we prepared for the Kuwait event, David's mother, Maggie, came to help the girls dress. She was an accomplished artist, she was mentored by Lauren Harris when she trained at the Ontario College of Art and completed a degree in cultural anthropology from La Universidad de San Carlos de Guatemala. Her father, a provincial archivist, had been a staunch supporter of the CCF, and her husband had been a senior diplomat. She laughed when she saw the invitation.

"My God! You're being invited to *'An Evening in the Garden of Earthly Delights.'* The Kuwaitis haven't learned a thing about Hieronymus Bosch."

We greatly enjoyed the event. While everyone else listened to the opening speeches, we explored the food spread and viewed most of the Kuwaiti treasures on display.

Neither Manna, Yuka, nor I expected to see the aged parliamentary assistant to the minister of health and welfare.

"These can't be your daughters!"

"Why not?" I asked.

I didn't want to profile Yuka's status as an emancipated minor and sensed a hint of racism, a microaggression.

Manna saved the day by putting out her hand.

"Manna Miwa-Garrick," she said, briefly curtseying. "This is my friend, Yuka Ichihara, who is visiting from Tokyo."

Yuka, who wore a crocheted cap and a sailor suit, closed her eyes and respectfully bowed. The ministerial assistant was charmed.

Then Jacinthe barged onto the scene, greeting me with kisses on both cheeks. Manna again put out her hand and curtsied, while Yuka bowed respectfully. An intrusive stare, followed by prolonged silence from Jacinthe and the gentleman accompanying her, managed to scare away the ministerial assistant. When he was out of earshot, she smiled broadly.

"I'm so glad you're here!"

She waved the ring finger on her left hand and displayed a large, yellow pear-shaped diamond.

"Lookie, lookie! There's going to be a wedding. And there are a couple of people here you should meet or reconnect with. They're both catching the end of the speeches and, of course, networking."

Yuka studied Jacinthe and remarked, "We practice piranha yoga. How often do you think we come across pomegranate juice?"

"And very acceptable smoked salmon," Manna added, "a quite tasty, sweet-olive tapenade, all very good for the digestive system.

"Well, it beats cod live oil," Jacinthe said, grimacing at the thought.

An official photographer approached and asked to take a picture.

"Oh, not me!" Jacinthe said. "Shoot them, they're such a lovely family. I'll go and rustle up the Leenies."

After the photo, two stunningly attractive women approached, and I immediately recognized Eileen Kathleen. Her hair, curlier than I remembered, was attractively arranged with tendrils framing her face. She wore a classic black cocktail dress and a large, intricately carved gold Haida bracelet on her right wrist. She also kissed me on both cheeks, and as I leaned in to return the gesture, I noticed her earrings, the unmistakeable six-point lines of light running across black star sapphires. They were not small. Both girls were delighted when I introduced Eileen Kathleen as my childhood friend.

"From Crofton?" Manna asked?

"Yes, indeed. And another childhood friend, Dr. Leenie Sprezzatoura."

Selina. I froze in place as a tall woman stood in the near distance. She wore a silk couture dress-and-jacket ensemble in a muted grey-beige colour, the rich fabric adorned with a pattern of olive-green discs, each bearing a printed image of a sun-god medallion. The jacket featured a pleated design and a Nehru collar. She had the presence of someone who could walk into, or out of, any palace in the world.

Her hair was cut stylishly short, and she wore green glittering peridot earrings. Her green snakeskin clutch purse contrasted with her crocodile heels and belt, expensive and tasteful."

"Selina!" I said, extending my hand.

She smiled and, as she hugged me, whispered, "I never use Selina, too much baggage comes with the name. Plus, one of the unfortunate young women in *Peyton Place* was named Selina. I prefer Leenie, and as a gesture of solidarity, when I officially changed my name, Eileen Kathleen changed hers too. Let's promise to keep in touch. We have so much to catch up on. And I assume we'll see you at the wedding."

Eileen Kathleen turned to the girls, then looked at me. "Your daughters?"

It had worked so effectively with the bureaucrat and, since Manna likes a bit of theatrics, she offered her hand, first to Eileen Kathleen, then to Selina, curtseying as she spoke her first name to each. Yuka, too, appreciated a touch of dramaturgy, repeating her bow to both of the Leenies before offering me a subtle nod and a warm smile.

"Our beloved guardian *ad litem*," she said. "Jeremy coparents us with Manna's father."

Lowering her voice, Yuka glanced around to see if anyone was within earshot before adding, "I am an emancipated minor."

After a few minutes of chit chat, Eileen Kathleen asked, "Have you had enough to eat? Let's go somewhere to catch up without the town's prying eyes and big ears imposing on our reunion."

The acting director of the Radiation Protection Branch stood not far across the room, surrounded by a phalanx of men in similarly smart suits. As he caught my eye, I smiled and waved. Noticing the interaction, Eileen Kathleen took my arm, with Manna following closely behind. Selina took Yuka's hand, and as Eileen Kathleen said, "I know a perfect place we can go to in the market," I heard Selina say, "I knew Jeremy when he was your age, just before I, like you, became an emancipated minor."

As we ascended the escalator, I turned toward the acting director. Although his friends were still deep in conversation, he watched our animated procession. I waved again and smiled, and his face lit up as he waved back. Turning to acknowledge him once more felt like the right thing to do.

Another unexpected outcome of attending the Kuwait function was Manna's aunt, Deborah, arranging for me to serve as a body double during His Royal Highness Prince Charles's scheduled visit to the National Gallery.

Prince Charles was visiting the gallery to showcase a publication of his watercolours and explore the exhibits. One of the highlights was a large upstairs room capable of displaying any historical print or drawing by a Canadian artist.

The curator, assigned to escort His Royal Highness, explained the technology used to access prints and drawings.

"Do you have a favourite Canadian print or drawing we may have in long-term storage?"

"Yes," I replied, "Untitled Tree, by Emily Carr."

He frowned. "Untitled Tree? With such a title, that may be difficult to locate."

"It was the promotional poster for a 1971 travelling exhibition at Le Musee de Beaux Arts in Montreal," I explained, "also the ROM and, of course, the VAG."

"Well, that certainly makes this easier," he said as he keyed the information into the device he carried.

While we waited, a beige metal cabinet glided along a grooved track in the floor near where we stood.

"How are you aware of this exhibit?" he asked.

"My paternal grandmother knew Emily. I'm sure his Royal Highness will appreciate seeing this drawing given its significance in Canadian art."

The moving cabinet stopped, and he scanned several drawers. "Drawer three."

He pressed a single key, triggering the drawer to open. There it lay, the brown paper bag on which Emily had sketched, in charcoal, the monumental cedar, Untitled Tree.

"Thank you *so* much!" I said, almost shouting my delight. "As a British Columbian, as a Vancouver Islander . . ." Caught by a sudden wave of emotion, I lowered my voice. "I'm humbled to see this and so very appreciative of your efforts in exhibiting it."

"Thank *you*, Jeremy! You've been a great tour rehearsal body double."

***

Those were the main highlights of my summer intern experience in the House of Commons. The regular thirty-day investigation updates from the RCMP, sent in response to my official complaint, were much less satisfying; they were just form letters.

When I returned to BC, I moved in with a probation officer whom I later discovered had had a relationship with a man I once dated. We both still loved him, and we always will.

One night, before I received the concluding report of the official complaint, I received a call at home.

"Gabriel?" the male voice asked.

"Mr. Gabriel," I replied, "or, if it's a social call, Jeremy. Only my cleaning woman calls me Mr. Gabriel, and I always correct her, Jeremy."

"Aren't *you* smart-mouthed!"

"Who is this, and why are you calling?"

"I'm calling to ask you to drop the investigation."

"Which investigation? I'm involved in several, usually referred to as assessments, sometimes even comprehensive risk assessments." Which assessment are you concerned with?"

"It doesn't surprise me you're involved in several

investigations," the caller said, mimicking my voice. "Don't you realize that *careers* are on the line?"

"What?" I was genuinely surprised. "That's not the kind of assessments I do."

"Are you taping or tracing this call?"

"Why would I?"

"Because you are who you are." Then the line went dead.

For days afterwards, I reflected on the call. I audited each of my ongoing assessments to determine which one might cause a career to be put at risk. I couldn't identify any case where the stakes would warrant such a call.

In early June, over a year after I had launched the official complaint, I received the final report. Not surprisingly, the two francophone officers were unable to identify if or when such an incident had taken place. I went over the letter with a fine-tooth comb. After reflection and analysis, I responded in writing, copying the letter to a provincial cabinet minister friend. The letter was officially stamped and headlined with.

# CHAPTER 29

**Strictly private and confidential**

FOR THE EYES OF THE MINISTER ALONE

Chief Superintendent (MKMC) OIC

Administrative and Personnel AOD,

Royal Canadian Mounted Police "E" Division,

657 West 37th Avenue,

Vancouver, BC V5Z 1K6

Re.: Your File, 92E-01453

Dear Chief Superintendent,

I am writing to acknowledge and thank you for your letter of June 10 with respect to the Complaint that I was discriminated against by members of a specific detachment on the basis of sexual orientation, and that once, I was called to assist two officers with a homicidal, suicidal subject then left alone in a hotel room with this man while the officers took off.

Like yourself, I regret that investigators couldn't arrive at a more satisfactory conclusion. As well, your letter has raised more questions than it has answered.

I have not decided whether to pursue a review of the Investigation through contact with the RCMP Public Complaints Commission. I would also like to very clearly express to you and your staff that I do not hold your agency and a majority of its members in ill repute. As I have previously advised, my grandfather was a member of the Northwest Mounted Police, my uncle served with your security intelligence service, and another member of my extended family is currently a member of the RCMP.

As a trained and practicing social scientist, however, I would like to offer criticism that I hope is viewed as constructive. I understand that certain historic prejudices have, until recently, been given official expression in policy and procedure within various jurisdictions of the federal government. I believe the ethos of a long-term and ill-advised Privy Council project, "The Fruit Machine," erodes very slowly.

You, sir, and I both function within the institutional cultures of public agencies. Having also worked as a special assistant in Centre Block of the House of Commons, I am aware of documentation that illustrates how specific programmes were funded and constructed to systematically exclude homosexual men and lesbian women from positions of influence within the Public Service of Canada, the Canadian Forces, and the Diplomatic Corps.

A legacy of such programs is that many positions of influence in government and public institutions are inaccessible to capable and intelligent men and women. A contextual reality is that those who occupy positions of confidence within specific institutions of government still believe it is their responsibility to exclude, through whatever means necessary, gays and lesbians from service in leadership capacities in the public domain. That is but one feature of institutional homophobia. When an individual behaves in a homophobic manner, it represents a pathological disorder. When this behaviour is formally or informally sanctioned by an institutional culture and is engaged in by one or more individuals, it becomes, systemically and systematically, conspiratorial and criminal behaviour.

Without rancour, it is my concern that the process and methodology of the investigation at issue have dictated the conclusions arrived at. In addition, there is a specific lacuna in the Investigation. It involves information I provided in Ottawa concerning the identity of the second member who was at the hotel.

Although I can't spell his name, and probably mispronounce it, I very specifically indicated which particular suicide he attended, mentioning the name of the suicidal subject, "in confidence," and in confidence only because the late gentleman has a surviving child who could be embarrassed if at any time, specific information about her father's death became part of an accessible public record. If you can't locate the investigating member's notes on this matter, I assure you that I documented the matter in two files. Who missed that one?

There is a specific (and statutory) inaccuracy in the information provided by my one-time clinical supervisor. Also, whoever analyzed the accumulated data suspended judgment at least a couple of times. For example,

I did get paid for the call-out. Although my clinical supervisor denied the overtime, his immediate supervisor, the regional manager, authorized it. Who else would have, or could have, authorized it?

Before speaking to the regional manager about the issue, I consulted an officer with appropriate jurisdiction in my union. If the regional manager hadn't authorized payment, I was prepared to launch a grievance, and the union was prepared to back me. The union's perspective was:

"The Employer provided your unlisted phone number to the RCMP. They called you on the day in question, not as 'Jeremy, Nice Guy', you were called as a professional social worker, 'regularly in the employ of the Province of British Columbia and as Chairperson of your Local and Shop Steward, just call the Regional Manager and ignore your immediate Supervisor.'"

I also remain convinced that collaborative investigative methods, including action-oriented research, consciousness raising, focus groups, a literature review of relevant material, and a reasonably structured and appropriately analyzed qualitative interview, would have rendered substantially different conclusions. Maybe even the truth.

A couple of issues endure. I was advised by at least one person alleging to be RCMP personnel that the "kid" in the hotel had a mental health worker, and it was not possible to reach that person. Was a profile of the homicidal, suicidal subject ever shared with the staff of the relevant mental health centre?

The attending physician and I learned from the subject, when he and I finally got to the local hospital, that he worked a rather bizarre "split-shift" at the local Safeway, and that reality, in part, had contributed to his state of higher-order anxiety. Was the Safeway lead ever checked out with either store personnel or the trade union representing organized employees at that closed shop?"

It appears there is a contradiction involving the information about a specific constable's capacity to be involved in the hotel incident. If he was on Municipal Traffic ". . . and would not normally be dispatched to such complaints," . . . how is it this specific Constable ". . . advised that he attended a similar type of incident with you." For the record, there was no similar incident in my entire career as a social worker, and there has been no similar incident in my entire life.

Further, I require clarification on the whole discourse about how, "... he remained at the scene for some thirty minutes. Things appeared to be calm, and he advised you that he would be departing the scene. He assured you, however, that if circumstances changed and the client began to show signs of distress or aggressiveness, you were to call the police, and they would re-attend immediately."

If it was the incident that I am concerned with, and not a "similar incident," when did these public-spirited assurances take place? When the man in the room was telling me he was ". . . sick and tired of cleaning up women who proudly describe themselves as lummoxes;" when he explained why he was ". . . sick of women who slept with multiple men;" perhaps when he described how "... nice girls and virgins just looked (at him), began to whimper, and ran out of the room.

Did the Constable intervene in another situation when someone was telling me he was "sick and tired of fucking (gerund) women," and he "... might as well turn gay right now?" I think not.

The referenced Constable was never that civil with me. That no one has been able to come up with any paperwork on the "similar type of incident" where this Constable was allegedly helpful and professional surprises me not at all.

I am prepared to acknowledge that this Constable may not have been the officer in charge of the intervention involving the male, mental health client at the hotel. I genuinely believed he was involved, but it may have been a case of mistaken identity. If that is the case, I am prepared to unreservedly apologize for any embarrassment or inconvenience he has experienced because of the anxiety I continue to feel about that unfortunate Sunday afternoon. I am reasonably certain, however, that one of the officers was French Canadian and the other was of Ukrainian Canadian origin.

As you are aware, I have discussed my major concern with academics, colleagues, members of the RCMP, and friends. A concern endures: If only the attending physician and I remember the subject, to what extent was the entire intervention socially constructed? Is there any possibility that the male subject, allegedly a mental health client, a victim of iatrogenic medical experimentation, a sexual assault victim, a parent at the age of twelve, an employee of Safeway, someone who presented with severe drug and alcohol dependencies, could have been a member of the RCMP and that the whole event was staged, got out of hand, and was rigorously supressed?

Any investigative procedure, or process, that continues to

deny my (shared) reality and (shared recollections) of the subject in question will only further cause me to believe that public institutions operating in northern British Columbia are blinded and incapacitated by a pervasive climate of homophobia, an environment not dissimilar from the cultural and institutional realities of race relations in various southern American States.

I do not doubt that homophobia and heterosexist discrimination exist in my Ministry. I also remain unconvinced that such vile sentiments and socially exclusionary practices and behaviours don't exist in your organization. I also believe that a frank, satisfactory resolve to my Complaint will go a long way to identify and address homophobia in both of our institutions.

Should you require any further clarification with respect to the concerns documented in this letter, I would look forward to assisting you or any member of your staff. Please don't hesitate to contact me during regular working hours at (Private line listed.)

Respectfully,

*Originally signed by the author*

Social Program Officer/ Senior Social Work Practitioner

Ministry of Social Services

Province of British Columbia.

(b)/copy: The Hon. Minister of Social Services, Province of British Columbia

***

# EPILOGUE

Nancy became a high-level child-welfare consultant and volunteered as a Search and Rescue co-ordinator in a Northern Vancouver Island community. She called to say that Carole-Anne had ALS and asked, if I would come to a 'Give-Away' of her belongings at a local recreation centre.

Although frail, Carole-Anne had organized various supporters to arrange all her belongings in the local recreation centre. She had a card printed to encourage whoever attended. If they financially contributed, they should let one of the women volunteers know, and they would get a tax receipt. Lots of people came. Most contributed for whatever they acquired. All the earrings went by noon. I left with Carole-Anne's copy of John Neihardt's book, Black Elk Speaks.

The night of the day she died, I attended the Victoria Labour Council monthly meeting. In the Good and Welfare section before the close of the meeting, feeling frail, I rose and informed the meeting that "With a heavy heart, I share that Sister Carole-Anne Dwyer passed away this morning."

As I spoke, for no more than the allotted three minutes for Good and Welfare announcements, I struggled to maintain my composure as I described how she was such a vigilant supporter of all Northern BC union members, frequently inviting them to stay in her home while she was off in other communities on union business.

As is our custom in the labour movement, when someone is at

the microphone and is struggling with an emotion-laden issue, another union member comes up and puts their arm around the person at the microphone. As I spoke, a young deputy sheriff put his arm around me.

Later, a Member of the Legislative Assembly rose in the House and commented on Carole-Anne's many contributions to the good and well-being of the Province's most vulnerable.

Later, Nancy gave me Carole-Anne's hulking silver Haida brooch, which a client had made for her. I had learned that early, during the Operation Solidarity mobilization against regressive legislation, Carole-Anne and a feminist colleague, a feisty Scots woman, climbed one of the major hills overlooking Prince George and, with a weed defoliant supplied by the Ministry of the Environment, wrote:

**Solidarity**

Across the slope. As the mobilization continued throughout the spring, the sign became more prominent and signalled to all that the workers were serious about opposing the legislation. Years later, I located the Scots woman, and on the floor of the Federation of Labour convention, I gave her the brooch. She touched her heart and said, "I will always honour the memory of Good Sister, Carole-Anne."

When I consulted Carole-Anne about how I was going to survive an oppressive and potentially toxic work environment, she asked, "Why do you want to commit your work-life to child protection?"

After some reflection, I let her know that, as someone who would never have children of my own, I was unwilling to be without children in my life. As someone sanctioned by training, legislation, and policy to be the legal guardian for those children whose families were unwilling or unable to care for them, I have cared for and advocated for vulnerable children who wouldn't be alive if I hadn't focused my knowledge, skill, experience, and attachment to their needs.

In an island village library, a noble woman audited her years of experience with social workers. ("The vice and the lice are still here.")

Our conversation was suddenly interrupted by a preteen. "Jeremy? Is it really you? I thought I'd never see you again!"

She thumped her chest, said her name, Druscilla, and mimicked a conversation we'd had in another community years ago.

"I'm doing well, not good. Mom's not drinking and she's working."

"Great."

"She made me this dress."

Druscilla wore a bright, polished cotton dress.

"Nineteen-fifties Simplicity pattern, back in style," she said. "For a while."

"Very smart," I replied.

"I'm impressed. You wear it well."

"Dad is in jail."

"That's too bad."

"No, it's not. Just breaches, not serious spit, but it'll teach him to keep from breachin', eh?"

A line from the Joan Baez song, *Marie Flore*, comes to mind: "I have seen children with faces much wiser than time."

Knowing she was taking time from my meeting, Druscilla extended her hand for me to shake.

"So good seein' you, Jeremy." She winked and laughed, "You've been a good fairy for my whole family."

She turned and walked away, but before she left the library, she waved, laughed again, pressed her hand to her lips, and blew me a kiss. Then, placing her hand over her heart, she nodded in a gesture of respect and walked away, still laughing.

### End

Et ignotas animum dimittit en artes *

* Literal translation: And he bent his mind to unknown arts. Inclusive-language translation, (imperative tense implied): Bend your minds to unknown arts.

Ovid Metamorphoses, VIII, 188

# Acknowledgements

This auto-ethnographic novel would not have occurred if I didn't survive the occupational safety deficits of inter-agency harassment and bullying within the workplace. I will be forever indebted to my friend, colleague and shop steward, Nancy Dwyer and her sister, the late Carole Anne Dwyer, who provided advice as a trade union executive and as a counsellor, that I document my early life experiences to re-discover what qualities of character and strategy assisted in my survival as a gay youth in the generally non supportive era of the early 1960s in a small community on Vancouver Island.

I am also appreciative of my union sisters and brothers in the Peace River and District Labour Council, particularly Ginger Lumin and Arla Lennox who provided consistent support and Jane Klem, who consistently affirmed the importance of language, linguistic scholarship and the value of tertiary education.

I am also appreciative of the work done by the late Lillian York and members of the Book Committee of the South Peace Arts Council who with the assistance of the Minister of National Health and Welfare and the New Horizons Program produced The Lure of the South Peace: From Birth to 1914, and the late Dorothy Calverly, historian, who documented the early history of white settlement in the Peace River area and whose work I have transliterated to provide context for the community of 'Dodge,' in the novel. As well, I will be forever indebted to Betty Hillman "and the Gallery Gang" for providing an autographed copy of Lillian's book project and a water colour by Lillian of the Rolla United Church which I donated to the South Peace Art Gallery and which now hangs in the Dawson Creek City Hall. And to Tim Hillman who took me to see Gregory Corso.

I am indebted to the work done by my cousins, Muriel Gabriel Heltzel and Marjorie Gabriel Neilsen in documenting the remarkable community service accomplishments and awards earned by their mother Sara Katherine Hitt Gabriel, who among other citations and awards received a 'Certificate of Merit' from the Unemployed Citizens League "for valuable service rendered in the interests of humanity and aiding the destitute and unemployed." (And for being welcomed at the White House by Eleanor Roosevelt for bringing linens made from flax grown in Oregon). I am always grateful for the friendship, advice and support of my fellow children's theatre performer, Judy Hill, of Judy Hill Galleries, who was always willing to listen, to clarify and endorse my perspectives of various community members who influenced the landed gentry characters who populate some chapters of the novel. Patricia Bacon-Rice was extraordinarily helpful in recalling many unique personalities - and usually supportive antics - of legions of activists who passed through the Earth Embassy during the mobilization to prevent mining uranium in British Columbia and during the organization of the April 26th International Day of Anti- Nuclear Protest. Similarly, I appreciate the efforts of Catherine Marie Gilbert, the author of the authorized biography of David Garrick, who when interviewing me about my friend and fellow 'eco-collaborator' reminded and re-sensitized me to the many contributions David made - for so long - to various sectors of the international and domestic environmental movement.

I am as well appreciative of the efforts of Professor Emerita, Celia Haig-Brown of York University, and author of Tsqelmucwilc: The Kamloops Indian Residential School -- Resistance and a Reckoning, who assisted me in clarifying the relationship between institutional ethnography and auto-ethnography and the development of these paradigms of analysis by the late Dr. Dorothy Smith.

I am particularly appreciative of the encouragement provided by my former practicum student Christine Woltman, BSW, MSW, RSW, who validated my recommendations of the importance of Strunk and White: The Elements of Style. I wish to acknowledge Irene Kavanagh, of Final Writes Editing and Writing Services who was recommended through contacts within the Chuckanut Writers' Conference. Finally, I want to acknowledge and thank Barnes Nobel Project Manager, Hazel Carter and her editorial team for her courtesy, patience and good humour in response to my perspective of the editorial process and my defence of "Canadianisms!" And Zayn Cole, Senior Author Strategist at Barnes Nobel for seeing and supporting the unique heroism and clarifying mission of No Backup - A Set-up and More: The Accomplishments of a Good Fairy.

I am indebted to three retired RCMP members who frankly discussed with me their experiences concerning negotiating their marriage plans with the Force. As well as being required to wait a probationary period before they could marry, their intended spouses needed to be approved. One retired member described how his wife, an accomplished cultural worker, and her family of origin who were actively involved with the Co-operative Commonwealth Federation (CCF) were initially subjected to significant scrutiny while the suitability of the bride - and the Force's ultimate approval of the marriage - was being assessed.

# FOOTNOTES

---

[1] Time to Be in Earnest: A Fragment of Autobiography, Prologue xiii. Faber and Faber Ltd. Bloomsbury House, London, 1999.

[2] "Clean for Gene," A slogan used in the 1968 Eugene McCarthy presidential campaign for long-haired anti-Vietnam War activists who cleaned up and dressed in a preppy manner to canvass for delegate support within the American Democratic Party.

[3] Shakespeare, William. Sonnet 18 'Shall I compare thee to a summer's day?' includes the verse: "Sometimes too hot the eye of Heaven shines."

[4] Proverbs 27-7

[5] T.S. Eliot, Four Quartets

[6] Coleridge, S. T. (1997). *Kubla Khan.* In W. Keach (Ed.), *The Complete Poems of Samuel Taylor Coleridge* (pp. lines 1–11). London: Penguin Classics.

[7] Shelley, P. B. (1904). *Ozymandias.* In T. Hutchinson (Ed.), *The Complete Poetical Works of Percy Bysshe Shelley* (lines 1–14). Oxford: Oxford University Press.

[8] The Gabriel-Hitt Family, compiled by Muriel Gabriel Heltzel and Marjorie Gabriel Neilsen, Library of Congress Catalog Card Number: 86-80833

[9] Ibid. p. 96.

[10] A biography of Wallis Simpson published in 2012 by Anne Sebba discussed possible sexual development disorder. At that time Dr. Norman Spack co-director of Gender Management Services at Children's Hospital, Boston challenged Sebba's argument saying "it is highly unlikely anyone will ever know if she was truly intersex."

[11] There is no proof that Wallis Simpson had an affair with Joachim von Ribbentrop though rumours have persisted for decades. Allegations suggest that she may have been romantically involved with him during his time as Germany's ambassador to Britain in the 1930's; these claims remain speculative and are not sustained by concrete evidence.

[12] "Sister Suffragette, music and lyrics by Richard M. Sherman and Robert B. Sherman for the film, Mary Poppins, Disney Studios, 1964."

[13] Then known as Aquat, now styled as Chusia.

www.ingramcontent.com/pod-product-compliance
Lightning Source LLC
Chambersburg PA
CBHW071318210326
41597CB00015B/1260